ACTIVITY BASED COSTING

THE KEY TO
WORLD CLASS PERFORMANCE

ACTIVITY BASED COSTING

THE KEY
TO
WORLD CLASS PERFORMANCE

Peter L. Grieco, Jr.

Mel Pilachowski

PT Publications, Inc.
4360 North Lake Blvd.
Palm Beach Gardens, FL 33410
(407) 624-0455

Library of Congress Cataloging in Publication Data

Grieco, Peter L., 1942–
        Activity based costing: the key to world class perfor-
mance/Peter L. Grieco, Jr., Mel Pilachowski.
                p.   cm.
        Includes bibliographical references and index.
        ISBN 0-945456-10-7: $29.95
        1. Cost accounting. I. Pilachowski, Mel, 1946–   .
        II. Title.
HF5686.C8G7845 1994
657' .42--dc20   94-846
                                                                          CIP

Copyright © 1995 by PT Publications, Inc.

Printed in the United States of America.

TABLE OF CONTENTS

PROFESSIONAL TEXTBOOKS

Available through PT Publications, Inc.
4360 North Lake Blvd.
Palm Beach Gardens, FL 33410

MADE IN AMERICA: *The Total Business Concept*
Peter L. Grieco, Jr. and Michael W. Gozzo

JUST-IN-TIME PURCHASING: *In Pursuit of Excellence*
Peter L. Grieco, Jr., Michael W. Gozzo and Jerry W. Claunch

SUPPLIER CERTIFICATION II: *A Handbook for Achieving Excellence through Continuous Improvement*
Peter L. Grieco, Jr.

BEHIND BARS: *Bar Coding Principles and Applications*
Peter L. Grieco, Jr., Michael W. Gozzo and C.J. (Chip) Long

SET-UP REDUCTION: *Saving Dollars with Common Sense*
Jerry W. Claunch and Philip D. Stang

WORLD CLASS: *Measuring Its Achievement*
Peter L. Grieco, Jr.

THE WORLD OF NEGOTIATIONS: *Never Being a Loser*
Peter L. Grieco, Jr. and Paul G. Hine

PEOPLE EMPOWERMENT: *Achieving Success from Involvement*
Michael W. Gozzo and Wayne L. Douchkoff

VIDEO EDUCATION SERIES

SUPPLIER CERTIFICATION: *The Path to Excellence*
A nine-tape series on World Class Supplier Based Management.

PREFACE

As we were writing this book, we kept having the image of a weed-infested lawn in front of a nice home. Every week, the owner would come out and mow down all the weeds. For about a day or so, the lawn looked pretty good. But after that, the weeds started popping up again all over the place.

In our minds, that is the state of costing in organizations today. We strongly believe the following are true:

- Traditional accounting practices are weakening American business.

- The emphasis on financial measurements diverts us from improvement.

- Traditional cost systems hinder excellence by hiding the elements of cost.

In other words, we mow down our problems every week and pretend that everything looks fine. And it does, for a couple of days. Then, those problems start popping up again all over the place.

Avoidance is not the answer when it comes to cost systems. If we do not search out the cost drivers in our companies and assign them directly to our products and services, we are going to lose out in the global marketplace.

We must get down on our hands and knees and pull the problems out by their roots. We must begin to use Activity Based Costing

(ABC) as a tool to find the areas of opportunity in our organizations. This is true whether we manufacture a product or provide a service. Our present way of assigning costs often makes problems worse.

ACTIVITY BASED COSTING shows you how to develop and implement a costing system that will provide you with the information you need to run your company in the most efficient and profitable manner possible. ABC is a natural partner for all of the latest business philosophies. It works exceptionally well in conjunction with Just-In-Time, Supply Based Management, Total Quality Management and Agile Manufacturing. In fact, ABC is the link which brings all of these techniques together under a single financial umbrella.

Each chapter of our book focuses on an important facet of the new costing paradigm. We have written it to provide you with information that will make you think about your present costing system and its limitations. But to stop there would be to stop short of our goal. Going beyond theory, our book gives you step-by-step examples of how to find the true Total Cost of producing a part or providing a service. Our book explores all areas of the company as well, not just the Finance or Accounting Department. We know that the war against weeds is fought on the shop floor.

This is an action book! Please read it and take the necessary steps to ensure your organization's prosperity and perhaps even its survival.

Peter L. Grieco, Jr.
Mel Pilachowski

Palm Beach Gardens, FL

ACKNOWLEDGMENTS

We would like to thank all of our clients for their good ideas, common sense and courage which was used to strive for continuous improvement. This book is their story of meeting one of America's most serious challenges. Special thanks go out to all of our colleagues at Pro-Tech who have challenged us and contributed their stories to this book. We thank them all for the time they took to review each chapter and make suggestions.

A special mention is reserved for the capable and hard-working office staff who are always there for us as we travel across the country and even to foreign countries. Much appreciation is due to Steven Marks for his creative editorial assistance. We wish also to thank Kevin Grieco for his design of our book cover.

We at Pro-Tech would also like to acknowledge in advance all the people who use this book and its ideas to bring their organizations into the twenty-first century. These people command our respect for their tireless efforts to bring costs under control and to make their companies and facilities into World Class institutions.

ACTIVITY BASED COSTING

THE KEY TO
WORLD CLASS PERFORMANCE

Chapter One

ACCOUNTING IS WEAKENING AMERICAN BUSINESS

Imagine a baseball game where the scorekeepers use complicated procedures and formulas involving players' salaries, batting averages, ticket sales, number of hot dogs sold during the third inning and projected draft picks to determine who won. They completely ignore the number of runs produced by each team. Imagine this scenario and you have an idea of what is going on in American businesses as accounting departments try to determine what companies have spent to build products or provide services. As we shall see, accounting results have little to do with the activities that take place on the factory floor, in the administrative offices or out in the sales field. Furthermore, the results we now use may be causing us to sink even deeper into a hole where we are no longer profitable or competitive with other companies in the global marketplace.

The traditional practice of accounting leads to a number of faults which end up weakening American business. Some of the more devastating are listed below.

- Traditional accounting practices fail to provide management with a window on performance because they focus on external (financial) results rather than on internal results.

- Traditional accounting practices reinforce tendencies by management to become increasingly more concerned with the bottom line instead of the production line.

- Traditional accounting practices fail to pinpoint troubles which are eating away at profit margins and providing an opportunity for the competition to capture market share.

- Traditional accounting practices fail to identify recovery opportunities in a company. They also overlook the real (root) causes of waste as well as cost reduction through the elimination of nonvalue-added activities or the improvement of value-added activities.

- Traditional accounting practices favor quick fixes to reduce costs either by reducing labor or by forcing suppliers to cut costs. These actions not only avoid internal issues of waste, continuous improvement, and the identification of true cost drivers and misallocated costs, but can create even more intractable problems via the neglect.

Avoidance is a big issue here and it reminds us of weeds in our lawns. Cutting them off at the surface only temporarily hides the problem. To eliminate weeds, you need to get at the roots and pull them out of the ground. Only then can you produce a healthier lawn. The same thing applies to our businesses. We need the tools to get at the roots and we need a cost management system that lets us use these tools to their full advantage.

Just-In-Time, Total Quality Management, Statistical Process Control, Set-Up Reduction, Bar Coding, Team Building and Supplier Certification are a few of these new tools. In our consulting work, we have seen them help put companies back on their feet and on their way to prosperity. But, we have also noticed that the greatest success comes when these tools are used in the context of a total cost system which looks to continuously improve performance by eliminating waste and reducing costs. The principal method whereby this happens is to assign each product its true share of activity costs and not to lump all costs which are not directly traceable into overhead which is then inaccurately allocated to products based on direct labor costs. This is the way we used to do product costing, but it is no longer a viable method. Let's take a look at some manufacturing history to see why.

The History of Costs
in Manufacturing

Assigning costs to products has been a way of doing business since time immemorial. The principal cost elements are material, labor and overhead. Labor and material have been allocated directly, that is, the cost of steel or a welder's hourly wage has been directly assessed against the product which was being built. All other costs, such as electricity, storage, insurance, benefits, rents, scrap and so on are dumped into the "catchall" barrel of overhead. Overhead costs are then allocated to units of production based on a predetermined burden rate which was correlated to labor dollars.

This system worked fine up through the 1970s when it started to become more and more clear that the figures weren't reflecting actual costs. Companies up to this point had been able to hide from these problems because price increases were allowing them to maintain profit margins, as were advances in technology and market share. In short, there was no need for aggressive cost management in prosperous times. But, cost systems were not keeping pace with the increasing complexity of operations, R&D advancements, multiple products, repetitive manufacturing and automation, to name a few areas. In addition, accounting had been increasingly turning toward the reporting of financial or external information in order to satisfy increasing public ownership of their companies and federal laws and regulations. It got to the point where reporting external financials was seen as the principal job of accounting, not reporting on internal costs and seeking ways to reduce them.

Another important shift that has occurred over the decades of this century is that labor costs began to constitute a smaller and smaller percentage of total costs as the chart on the next page shows:

CHANGING COST BEHAVIOR PATTERNS

Today		In 5 Years
25-45%	OVERHEAD	19-25%
		0-6%
5-25%	DIRECT LABOR	75%
50%	MATERIALS	

As we have noted, however, overhead and other indirect costs were allocated to products based on labor dollars when labor costs were a much more significant factor. Unfortunately, the old cost system still operates under the assumptions of the late 19th and early 20th century when it made sense to allocate overhead costs based on labor. This way of doing business can no longer continue. Product cost data must be made to reflect the real activities that are performed while making the product. And cost systems need to give managers the information they really need to make intelligent decisions.

The Growing Dissatisfaction with Managerial Accounting

Providing management with the information it needs to make informed, rational decisions is the job of managerial accounting.

Obviously, from the preceding discussion, there are a great many accounting practices which no longer give management an accurate picture of what is truly happening in their company. There is reality and then there are the numbers and reports that the accounting department issues. Dissatisfaction with managerial accounting groups itself into the following areas:

- **Crude costing methods distort product costs.**

- **Wrong sourcing decisions.**

- **Poor investment analysis.**

- **Internal information not timely or useful because of emphasis on external reporting.**

- **Return on Assets (ROA) based on book value, not real worth.**

- **Excessive emphasis on direct labor.**

- **Short-term results minimized at expense of long-term profitability.**

- **Cost and manufacturing systems are not integrated.**

- **Lack of cooperation between departments.**

This growing dissatisfaction with managerial accounting is the result of changing conditions in the business world. As we all know, direct labor is a decreasing percentage of manufacturing costs while indirect costs are an increasing percentage. Unfortunately, traditional cost systems do not reflect that change. Thus, they target labor for cost reduction because that is the way it was always done and because labor is the most visible element of cost. But indirect costs are what really needs to be looked at, as we shall soon discover.

There have also been changes in the relevance of material costs in the new business climate. It has traditionally been considered the easiest cost element to control, but that advantage is slipping away because the data which is collected, controlled and reported on material costs is too often inaccurate. In addition, it is no longer as effective because it is based on cost data which is growing obsolete. Purchase Price Variance (PPV) and material overhead, for example, do not represent the true cost of procuring, receiving and stocking material, nor are outside supplier's operations applied directly to product cost. The result is a distorted picture of true product cost. Products are being assessed cost from overhead based on irrelevant factors instead of actual usage. Granted, the methodology used to allocate overhead in this manner is easier, but it lacks any degree of accuracy and, thus, usefulness.

The example given here shows how and why the traditional method fails.

Example 1

Traditional Method

Product A consists of **20 components** costing **$180** and has an assembly time of **2 hours** at a labor cost of **$20**. Overhead is applied based on a direct labor factor of **120%** which results in a cost of **$44**. The total cost of **Product A** is **$244**.

> **Product B** consists of **5 components** costing **$180** and has an assembly time of **2 hours** at a labor cost of **$20**. Overhead is applied based on a direct labor factor of **120%** which results in a cost of **$44**. The total cost of **Product B** is **$244**.

If you were to look at this comparison more deeply, however, you would begin to question whether the amount of overhead utilized to produce both products was the same. But if we do look more closely, we see that there is a substantial difference between the two products in the number of components. Logic tells us that the cost associated with assembling 20 components must be more than the cost associated with assembling 5 components. Indirect overhead costs — ordering, receiving, stocking and issuing — are not properly accounted for if we just use direct labor as the factor by which we allocate overhead. This is the way in which traditional costing methods distort true product cost.

The calculation of overhead points to yet another change that we must bring to our costing systems. Traditional cost management systems lumped all nonvalue-added activities into overhead. From this lump sum, all products are charged regardless of the product's use of the activity center. The calculation and aggregate reporting of overhead has hidden its true cost relationship to products. Some companies have begun to move in the right direction by increasing the number of overhead rates in order to minimize product cost distortions. The use of a single overhead rate should be discontinued. Cost allocations should be based on an appropriate *cause* and *effect* relationship between cost and the activities which are used.

It should be clear that we need to change our mindset about managerial accounting. With the advances in manufacturing philosophies and the introduction of a new set of challenges, there is no other choice but change. Repetitive and flexible manufacturing, Just-In-Time, plants within plants, and Set-Up Reduction are leading the way and cost management is still lagging behind. Why is this so? There are five reasons why we think this growing split between managerial needs and accounting practices has developed:

- **Insufficient knowledge base.**

- **Issues of practicality.**

- **External factors.**

- **Conservative bias.**

- **Manufacturing complexity.**

In our book, we aim to explore each of these deficiencies in our discussions of cost management topics. As H. Thomas Johnson and Robert S. Kaplan have shown in their book, *Relevance Lost: The Rise and Fall of Management Accounting*, accounting systems are out of touch with a proactive management's needs in today's global marketplace. We need to spend less time measuring the bottom line and more time controlling the production line.

Accounting's Failure to Provide a Basis for Performance Measurement

One of the most troubling failures of accounting is its focus on the measurement of external factors instead of concentrating on the measurement of internal activities. Perhaps the most common complaint about the number crunchers and bean counters comes in the form of the question below:

> # What do accounting reports
> # have in common
> # with manufacturing results?

The sad reply in many companies today is that they have nothing in common. How can we run businesses when these two vital components are not in synch? The answer is that many of us can't do so profitably, as evidenced by our trouble competing in global markets. And, until we shift our emphasis away from developing measurements which satisfy financial reporting requirements and toward measurements which provide management and workers with information to use for improvement, we will continue to flounder. In fact, the gulf between what we need and what we get has led many of us to ignore the performance measurements which accounting uses. Needless to say, this is not a healthy situation either.

Because of the emphasis today on financial measurements, we are given aggregate figures which do not promote improvement or the identification of cost drivers, or waste. A departmental scrap report, for example, does not detail specific areas to be investigated. Instead, it alarms management whose reaction is to order a decrease in scrap without offering any concrete suggestions for eliminating waste. That's nobody's fault. The present system of assigning costs provides us with no idea of why scrap levels are high. A total cost system, on the other hand, has the primary purpose of identifying sources of scrap and then attacks them by pulling them out at the roots, as we mentioned earlier. A total cost system, in fact, measures the activities which contribute to scrap and seeks to drive them out instead of accepting them as a given and throwing them into the catchall category of overhead. We need to produce reports that identify operators, machines, parts or suppliers who are in need of improvement so that we can concentrate on solutions.

The challenge for cost systems in an agile environment is to provide us with performance measurements which help us meet the following business objectives:

AGILE MANUFACTURING
<u>BUSINESS OBJECTIVES</u>

- **Lower inventories.**

- **Lower product cost.**

- **Smaller lots.**

- **Improved quality.**

- **Decreased lead times.**

- **Increased productivity.**

- **Improved customer satisfaction.**

We can meet these objectives by identifying cost drivers, some of which are obvious and others which are hidden, and their impact on total product cost. That done, we then need to control and monitor these factors which contribute to waste. Management must come up with performance measurements which are based on these objectives and which turn away from traditional measurements and reports. We need to investigate and develop measurements which integrate the organizational structures and manufacturing activities of our companies and which make the entire company aware of the need to ferret out the hidden elements of cost.

Manufacturing Excellence Hindered by Hiding the Elements of Cost

As we have already pointed out, traditional cost systems put most of their effort into capturing material and labor costs. All other costs are relegated to overhead accounts. At the same time, management has come to accept overhead as a given and believe that there is not much a manager can do to improve upon this cost area. In the minds of traditionalists, it is the cost of doing business. Therefore, the great majority of effort is put into improving upon labor and material costs. There are two things wrong with this reasoning.

First, traditional cost systems rely on far too few elements of product cost to paint a realistic picture. Second, this reliance upon three elements of cost (labor, material and overhead) hides the real opportunities for improvement. In order to rectify this situation, we need to restructure our system of reporting costs so that it monitors and controls many more elements and identifies these elements back to the correct product and level. This will entail addressing the cost of a number of functional activities within

Manufacturing and beyond. We need systems which will report on the activities of Sales, Engineering, Finance and Procurement as they relate to each product.

In addition, we need to address the issue of direct and indirect costs. It is an issue which is central to uncovering the hidden elements of cost. As we all know, costs can be either directly or indirectly related to a particular item being produced. *A cost is not direct or indirect in and of itself, but only in relation to the given item being produced.* Since direct costs can be identified more clearly to the products manufactured, they are usually thought to be more easily controlled. On the other hand, indirect costs usually become a cost of the product only through allocation of overhead accounts by some predetermined formula, if they get allocated at all. Such a situation makes indirect costs very difficult to control.

The challenge of the future is to link manufacturing costs more directly to the products we build. If this is not possible, then we should be asking ourselves whether the activity can be eliminated entirely. We need to bring hidden costs into the open in order to provide ourselves with a more accurate picture. This level of accuracy will allow us to eliminate waste and thus improve quality, delivery and cost so we can become agile manufacturers. Only by taking these measures can we expect to achieve a total cost analysis which will provide management with the data it needs to make decisions.

Lack of Meaningful Data
to Support Management Decision Making

In an era where economic conditions, global competition and lower margins are making every business decision a critical one, the need for meaningful data increases exponentially. Here are just some of the important decisions we must face every day:

- **Make vs. buy.**

- **Sourcing.**

- **Price setting.**

- **Capital investing.**

- **Product mix.**

- **Inventory levels.**

- **Lot sizing.**

- **Product abandonment.**

A proper analysis of the above decisions cannot be made without accurate performance and cost data.

Unfortunately, American companies have been asked to make decisions based on data which is misleading and often inaccurate. Consequently, we have made bad decisions which have resulted in lost profits, market shares and even whole industries. When we rely on reports which measure labor efficiency, for example, we are obscuring the far more important issue of product quality. Purchase price variance which focuses on differences to the standards helps us avoid concentrating on the creation of supplier partnerships. In general, any job costing system which reflects performance against a standard steers us away from looking at the causes and effects of poor performance by putting the focus on the wrong kind of data.

Lack of meaningful data to support management decision-making is, we believe, at the core of the manufacturing issues we face. Companies buy more material than they need. They make more product than is required. They borrow more money than they want. They invest in more square footage, machines and people instead of investing in the process of improvement. This is the key component of our message. Traditional approaches to product costing leads to decisions which cover up the manufacturing issues which are weakening our companies and even our economy. We should stop spending money on systems and ideas which don't place their emphasis on continuous improvement.

Given the proper cost reports, management could start to focus on investment in the reduction or elimination of nonvalue-added activities instead of increasing in size or bureaucratic complexity. Our organizations are getting fat with large inventories. Waste makes our facilities bulge at the seams. Sooner or later, we weigh ourselves and notice our unhealthy state of fitness. And then all too often we panic and start drastically slashing inventory, spending and people. We are suggesting in this book that you stop panicking. Our new weighing scales don't advocate crash diets or liposuction. Instead, they suggest that you find a "doctor" who can help you get on a diet which will show you the most productive way to reduce. We need a job costing system which reduces fat and waste, but allows our companies to maintain a healthy and attractive financial appearance. This will entail looking at more than the traditional costing data.

Activity Based Costing

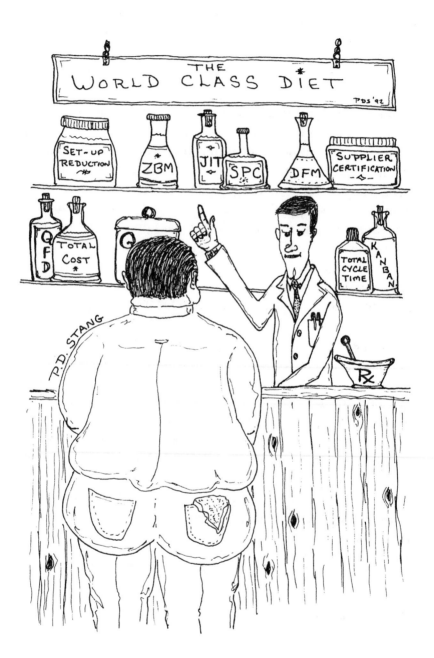

Absence of Nonfinancial Data
when Reporting Results

Cost systems around the world focus on financial issues which relate directly to dollars spent or to dollars earned. We have seen this practice, although accepted and auditable, restrict the process of continuous improvement. One of the reasons why the traditional systems fail is that they fail to report and control nonfinancial cost measures, costs which can expose the cause and effect behind the activity presently being performed. For example, we look at the labor, material and overhead costs of building a product, but we rarely ask ourselves what it costs our company for the following wasteful activities.

WHAT DOES IT COST . . .

- **For a customer to complain?**

- **For materials to sit?**

- **To miss a delivery?**

- **To process a change order?**

- **To enter a transaction?**

- **To complete a time card?**

- **To process a sales order?**

These activities are not normally considered in cost results. However, their existence causes dollars to be spent on wasteful activities. The effect of these wasteful dollars can be expensive, even though traditional systems don't capture them. In fact, they avoid them.

A new product costing system must start identifying these nonfinancial costs, which are mostly waste, and then help other parts of the company to eliminate or reduce them as much as possible. We have found that simply drawing attention to these issues and their associated costs is often enough to motivate a company to want and seek improvement. Nonfinancial costs are usually high and therefore offer big opportunities for savings. In one company we work with, for example, the cost of a customer to complain was analyzed to cost $850 which did not include the replacement cost of a new product. In doing this analysis, the company also captured the number of complaints which were processed during the month in order to compute the total cost of customer complaints. Here is what they found:

Number of Complaints		
X **Complaint $s** = **Total Cost**		
		of Customer
		Complaints
7 X $850	= $5,950	

The analysis soon showed the company that the dollars associated with this nonfinancial activity were large enough to track. The company began to report complaints over a period of time in order to generate an average dollars figure to use as a nonfinancial result. The next step, of course, will be to find ways to reduce both the number of complaints and the cost of responding to a customer complaint.

Summary

Product cost systems need to change. Companies are facing a new set of challenges in the age of automation and global competition. The traditional systems are hiding some of the costs we need to know in order to improve. Our responsibility is to get out of our offices and into the operations of our companies. We must understand how our company builds its products so that we can begin the task of monitoring and controlling activities. *This attention to the production line will inevitably result in savings to the bottom line.*

Chapter Two

RECOGNIZING THE CHANGES REQUIRED

Horses wear blinders so that they don't get distracted by all the activity going on around them. That way, they will keep focused on one direction no matter what else may be happening around them. It is exactly for that reason that companies should stop wearing blinders. Companies with blinders don't know what is going on in the marketplace that surrounds them; they don't even know what is actually happening in their own companies. Like the horses, they slowly shuffle along the same path if they are lucky. If they aren't, these "blinded" companies simply cease to exist.

Our current cost structure is like a pair of blinders. It prevents our companies from recognizing the changes we need to keep pace and feeds us information that does not help us achieve strategic objectives. Costs are never, or rarely, directly traceable to performance of the products we build. And without knowing what our total costs are for a particular product, how can we ever support cost reduction? When the results of collecting cost information are neither within predictable bounds or accurate, then we have little information upon which to base improvement efforts. Instead of problem solving in such an environment, we perform symptom solving.

It is this concern with symptom solving based on an inaccurate cost structure that leads us to ignore the major issues which confront our companies. The direct results are often decreasing profit margins and market shares. And that is no surprise, because how is it possible for a company to improve if it fails to discover the opportunities within a company's activities for cost improvement. But where should a company start its improvement efforts? We believe you should start with the company's strategic plans. And that takes more than just saying that we want to improve quality, satisfy customers, reduce costs or increase inventory turns. All those endeavors are good ones, but make sure that you aren't just blowing smoke because that is what you are supposed to say. Companies need to benchmark their current performance with cost data which points out areas where change is needed the most.

And, most importantly, we need to get up and initiate new activities and projects which actually start to solve our quality and cost problems. Improvement may begin with recognition of the problem and discussion of solutions, but nothing gets done until we *stop talking* and *start acting*. We can't put this any stronger.

You are not improving until you start acting. So, get out of your easy chair and get involved.

Measurements that Support Organizational Objectives

It is a well known fact that companies should be driven by organizational objectives. If these objectives meet the business need to be profitable and if people are focused on achieving these objectives, then most companies will be successful. Each company will, of course, have a different mix of objectives, but all companies in the future must have a means of measuring these objectives. This means of measurement must contain the ability to apply costs directly to each product in order to arrive at a total cost. Since most organizations around the world have the same objectives, let's look at two of the more important ones, Quality and Customer Satisfaction, and how we can measure them.

QUALITY — To measure quality, companies must start reflecting costs in terms of dollars and not just percentages. Mel Pilachowski sums it up in this way:

> **Quality percentage levels reflect how bad we are.**
> **Quality dollars reflect why.**

Furthermore, quality dollars should be measured at the conformance and nonconformance level. Today, companies primarily collect and report only on conformance dollars. But, quality is not just the responsibility of one or two areas of a company as is reflected in this type of cost accounting. That is why, when such companies set out to improve quality, they often fail. They are only making a partial effort and their measurement of conformance dollars only reports on a small part of the total picture. Thus, the results are disappointing at best and misleading at worst.

Total Quality Management (TQM) suggests that every person in a company is responsible for quality. There is more to an organization than just manufacturing. Both the cost of conformance and nonconformance need to be measured function by function against quality. Look at the forms we have included here to see just what needs to be measured.

COST OF QUALITY FORMS

Sales and Marketing
Cost of Quality Categories — Elements

Cost of Conformance	Cost per Occurrence	Number of Occurrences	Extended Cost
1. Procedures	_____	_____	_____
2. Training	_____	_____	_____
3. Forms Design	_____	_____	_____
4. Sales Support Material	_____	_____	_____
5. Design Specifications	_____	_____	_____
6. P&L	_____	_____	_____
7. Computer Data	_____	_____	_____
8. Market Forecast	_____	_____	_____
9. Legal and Product Safety Review	_____	_____	_____
10. User Market Research	_____	_____	_____
11. Sales Support Cast	_____	_____	_____
12. Customer Survey	_____	_____	_____
13. Sales Dollars	_____	_____	_____
14. Service Cost by Area/Advertising	_____	_____	_____
15. Loss Leaders	_____	_____	_____
16. Launch and Field Test	_____	_____	_____
17. Pilot and Field Test	_____	_____	_____
18. Incentive Programs	_____	_____	_____
19. Market Survey	_____	_____	_____
		Subtotal	_____

Sales and Marketing
Cost of Quality Categories — Elements
(continued)

Cost of Nonconformance	Cost per Occurrence	Number of Occurrences	Extended Cost
1. Labor of Redo's — Administration	_____	_____	_____
2. Incorrect Order Entry	_____	_____	_____
3. A/R Receivables	_____	_____	_____
4. Special Instructions	_____	_____	_____
5. Field Service — Excessive	_____	_____	_____
6. Warranty	_____	_____	_____
7. Literature Reprint	_____	_____	_____
8. Contingent Liability	_____	_____	_____
9. Unit Productivity	_____	_____	_____
10. T&E	_____	_____	_____
11. Product Recall	_____	_____	_____
12. Loss of Market Share	_____	_____	_____
		Subtotal	_____

Engineering
Cost of Quality Categories — Elements

Cost of Conformance	Cost per Occurrence	Number of Occurrences	Extended Cost
1. Design Specification Review			
2. Product Qualification, Evaluation, Characterization			
3. Drawing Checking			
4. Supplier Evaluation			
5. Preventive Maint.			
6. Process Capability Studies			
7. Fabrication of Special Test Fixtures			
8. Verify Workmanship Standards			
9. Review of Test Specifications			
10. Failure Effects/Mode Analysis			
11. Pilot Procedure Runs			
12. Packaging Qualifications			
13. Customer Interface			
14. Safety Review/Operator Safety			
15. Technical Manuals			
16. Preproduction Reviews			
17. Defect Prevention Program			
18. Schedule Reviews			
19. Process Reviews			
20. Early Approval of Product Specifications			
21. Computer Aided Design (CAD)			
22. First Piece Approval			
23. Agency Approval			

Engineering
Cost of Quality Categories — Elements
(continued)

24. Supplier Qualification	_____	_____	_____
25. Special Test Fixture Design Review	_____	_____	_____
26. Education	_____	_____	_____
27. Prototype Inspection and Test	_____	_____	_____
28. Testing	_____	_____	_____
29. Receiving Sample Testing	_____	_____	_____
30. In-Process Sample Testing	_____	_____	_____
31. Final Sample Testing	_____	_____	_____
32. Laboratory Analysis and Test	_____	_____	_____
33. Fault Insertion Test	_____	_____	_____
34. Engineering Audits	_____	_____	_____
35. Training for Special Testing	_____	_____	_____
36. Personnel Appraisal	_____	_____	_____
		Subtotal	_____

Cost of Nonconformance	Cost per Occurrence	Number of Occurrences	Extended Cost
1. Warranty Expense	_____	_____	_____
2. Engineering Travel and Time on Problems	_____	_____	_____
3. Engineering Change Notices	_____	_____	_____
4. Redesign	_____	_____	_____
5. Premium Freight Cost	_____	_____	_____
6. Material Review Activities	_____	_____	_____
7. Failure Analysis (Return Evaluation)	_____	_____	_____
8. Corrective Action	_____	_____	_____
9. Failure Approach	_____	_____	_____
10. Return Goods Analysis	_____	_____	_____
11. Product Liability (Design Related)	_____	_____	_____
		Subtotal	_____

Industrial Engineering
Cost of Quality Categories — Elements

Cost of Conformance	Cost per Occurrence	Number of Occurrences	Extended Cost
1. Operator Training			
2. Design Review			
3. Inventory Control			
4. Job Description			
5. Methods Description			
6. Test Equipment Description Verification			
7. Material Utilization			
8. Line Rebalance			
9. Process Verification			
10. Product Control Card System			
11. Material Usage Verification			
		Subtotal	

Cost of Nonconformance	Cost per Occurrence	Number of Occurrences	Extended Cost
1. Tool Repair			
2. Tool Modification			
3. Corrective Action Costs			
4. Engineering Change Order			
5. Purchasing Change Order			
6. Turnover			
7. Obsolete Job Description			
		Subtotal	

Procurement and Acquisition
Cost of Quality Categories — Elements

Cost of Conformance	Cost per Occurrence	Number of Occurences	Extended Cost
1. Supplier Review and Approval	_____	_____	_____
2. Send Proper Specifications to Supplier — Make it Clear What is Required	_____	_____	_____
3. Periodic Seminars	_____	_____	_____
4. Forecasting — Cost of Carrying Hard-to-Get Materials	_____	_____	_____
5. Material Cost Resulting from Multiple Sourcing	_____	_____	_____
6. Strike Build-Up Costs	_____	_____	_____
7. Evaluation of Supplier's Equipment that Will be Used to do the Job	_____	_____	_____
8. Review Supplier Incoming Quality Practices	_____	_____	_____
9. Recertification of Suppliers	_____	_____	_____
10. Incoming Inspection Cost	_____	_____	_____
11. Information Systems Cost Associated with Supplier Rating	_____	_____	_____
		Subtotal	_____

Procurement and Acquisition
Cost of Quality Categories — Elements
(continued)

Cost of Nonconformance	Cost per Occurrence	Number of Occurrences	Extended Cost
1. Scrap	_____	_____	_____
2. Sorting	_____	_____	_____
3. Reinspection Due to Rejects	_____	_____	_____
4. Rework	_____	_____	_____
5. Excess Inventory Due to Lack of Confidence in Supplier Delivery	_____	_____	_____
6. Loss Incurred as a Result of Supplier Delinquencies	_____	_____	_____
7. Corrective Action Cost	_____	_____	_____
8. Shipping Cost on Returns to Suppliers	_____	_____	_____
9. Purchase Order Rewrite	_____	_____	_____
10. Purchase Order Changes Resulting from Error	_____	_____	_____
11. Premium Freight	_____	_____	_____
12. Trips to Suppliers to Resolve Problems	_____	_____	_____
13. Expediting Cost to Assure Proper Deliveries (i.e., Phone Bill)	_____	_____	_____
		Subtotal	_____

Plant Administration
Cost of Quality Categories — Elements

Cost of Conformance	Cost per Occurrence	Number of Occurrences	Extended Cost
1. Consultants	_____	_____	_____
2. Preventive Maintenance Program	_____	_____	_____
3. Modeling	_____	_____	_____
4. Controlled/Critical Storage	_____	_____	_____
5. Environmental Control	_____	_____	_____
6. Labor Training	_____	_____	_____
7. Review of Labor Production Rates	_____	_____	_____
8. Security	_____	_____	_____
9. Surveillance	_____	_____	_____
10. Machine Maintenance — P.M.	_____	_____	_____
11. Machine Maintenance — Training	_____	_____	_____
12. Timely Machine Replacement	_____	_____	_____
13. Equipment Depreciation Review	_____	_____	_____
14. Equipment Depreciation Reappraisal	_____	_____	_____
15. Facility Planning — Audits	_____	_____	_____
16. Facility Inspection and Testing	_____	_____	_____
17. Data on Labor Productivity	_____	_____	_____
18. Pilot Run to Check Standard	_____	_____	_____
19. Labor Surveillance	_____	_____	_____
20. Time Card Control Test	_____	_____	_____
21. Time Card Audit	_____	_____	_____
22. Machine Maintenance Test	_____	_____	_____
23. Machine Maintenance Inspection	_____	_____	_____

Plant Administration
Cost of Quality Categories — Elements
(continued)

24. Equipment Depreciation
    Inventory   _____   _____   _____
25. Equipment Depreciation
    Audit   _____   _____   _____
26. Equipment Depreciation
    Tracking   _____   _____   _____
                                    Subtotal _____

Cost of Nonconformance	Cost per Occurrence	Number of Occurrences	Extended Cost
1. Facility Planning Redesign	_____	_____	_____
2. Facility Equipment Replacement	_____	_____	_____
3. Missed Schedule	_____	_____	_____
4. Incorrect Labor Level	_____	_____	_____
5. Increased Failure	_____	_____	_____
6. Incorrect Time	_____	_____	_____
7. Machine Scrap	_____	_____	_____
8. Machine Rework	_____	_____	_____
9. Machine Downtime	_____	_____	_____
10. Product Liability	_____	_____	_____
11. Equipment Depreciation — Obsolete	_____	_____	_____
12. Equipment Depreciation — Premature	_____	_____	_____
		Subtotal	_____

Activity Based Costing

Quality Control
Cost of Quality Categories — Elements

Cost of Conformance	Cost per Occurrence	Number of Occurrences	Extended Cost
1. Quality Training	_____	_____	_____
2. Test Planning	_____	_____	_____
3. Inspection Planning	_____	_____	_____
4. Audit Planning	_____	_____	_____
5. Product Design Review	_____	_____	_____
6. Supplier Qualification	_____	_____	_____
7. Producibility/Quality Analysis Review	_____	_____	_____
8. Process Capability Studies	_____	_____	_____
9. Machine Capability Studies	_____	_____	_____
10. Calibration of Quality Equipment	_____	_____	_____
11. Operator Certification	_____	_____	_____
12. Incoming Inspection	_____	_____	_____
13. In-Process Inspection	_____	_____	_____
14. Final Product Inspection	_____	_____	_____
15. Product Test	_____	_____	_____
16. Product Audit	_____	_____	_____
17. Test Equipment	_____	_____	_____
18. Checking Gauges, Fixtures, etc.	_____	_____	_____
19. Prototype Inspection	_____	_____	_____
20. Quality Systems Audits	_____	_____	_____
21. Customer/Agency Audits	_____	_____	_____
22. Outside Lab Evaluations	_____	_____	_____
23. Life Testing	_____	_____	_____
24. Product Audits	_____	_____	_____
		Subtotal	_____

Quality Control
Cost of Quality Categories — Elements
(continued)

Cost of Nonconformance	Cost per Occurrence	Number of Occurrences	Extended Cost
1. Scrap Analysis	_____	_____	_____
2. Rework Analysis	_____	_____	_____
3. Warranty Cost Analysis	_____	_____	_____
4. Concessions Analysis	_____	_____	_____
5. Factory Returns Analysis	_____	_____	_____
6. Material Review Board Action	_____	_____	_____
		Subtotal	_____

Manufacturing
Cost of Quality Categories — Elements

Cost of Conformance	Cost per Occurrence	Number of Occurrences	Extended Cost
1. Training: Supervisor Hourly			
2. Special Review	_____	_____	_____
3. Tool/Equipment Control	_____	_____	_____
4. Preventive Maint.	_____	_____	_____
5. Zero Defect Program	_____	_____	_____
6. Identify Incorrect Specs/Drawings	_____	_____	_____
7. Housekeeping	_____	_____	_____
8. Controlled Overtime	_____	_____	_____
9. Checking Labor	_____	_____	_____
10. Trend Charting	_____	_____	_____
11. Customer Source Inspection	_____	_____	_____
12. First Piece Inspection	_____	_____	_____
13. Stock Audits	_____	_____	_____
14. Certification	_____	_____	_____
Subtotal			_____

Cost of Nonconformance	Cost per Occurrence	Number of Occurrences	Extended Cost
1. Rework	_____	_____	_____
2. Scrap	_____	_____	_____
3. Repair and Return Expenses	_____	_____	_____
4. Obsolescence	_____	_____	_____
5. Equipment/Facility Damage	_____	_____	_____
6. Repair Equipment Material	_____	_____	_____
7. Expense of Controllable Absence	_____	_____	_____
8. Supervision of Mfg. Failure Element	_____	_____	_____
9. Discipline Costs	_____	_____	_____
10. Lost Time Accidents	_____	_____	_____
11. Product Liability	_____	_____	_____
Subtotal			_____

Comptroller
Cost of Quality Categories — Elements

Cost of Conformance	Cost per Occurrence	Number of Occurrences	Extended Cost
1. Forecasting Perf.	_____	_____	_____
2. Training/Procedures	_____	_____	_____
3. Ledger Review of P&L and Balance Sheet	_____	_____	_____
4. Budget Generation	_____	_____	_____
5. Long Range Planning	_____	_____	_____
6. Job Description	_____	_____	_____
7. Cost of Quality Budget	_____	_____	_____
8. Time Card Review	_____	_____	_____
9. Capital Expenditure Review	_____	_____	_____
10. Total Expenditure Reviews/ Delegation of Authority	_____	_____	_____
11. Order Entry Review	_____	_____	_____
12. Product Cost Std.	_____	_____	_____
13. Cost Reduction	_____	_____	_____
14. Cost of Quality Review	_____	_____	_____
15. Data Processing Report/ Fin'l Report Reviews	_____	_____	_____
16. Ledger Reviews	_____	_____	_____
17. Invoicing Review	_____	_____	_____
		Subtotal	_____

Cost of Nonconformance	Cost per Occurrence	Number of Occurrences	Extended Cost
1. Billing Errors	_____	_____	_____
2. Inv. Out of Control	_____	_____	_____
3. Improper A/P Supplier Payments	_____	_____	_____
4. Incorrect Accounting Entries	_____	_____	_____
5. Bad Debts Turnovers, Overdue A/R	_____	_____	_____
6. Payroll Errors	_____	_____	_____
		Subtotal	_____

Activity Based Costing

Software
Cost of Quality Categories — Elements

Cost of Conformance	Cost per Occurrence	Number of Occurrences	Extended Cost
1. Software Planning			
2. Software Reliability Projection/Prediction			
3. Systems Analyst Interrogating Activities			
4. Documentation Review			
5. Learn/Understand Customer Requirement/ Business			
6. Preparation and Review of System Specifications			
7. Flow Chart Review			
8. Correlation Analysis			
9. Keypunch Operator Training			
10. Tape Duplication and Verification			
11. Program Testing			
12. Function Testing			
13. Performance Test			
14. Code Verification			
15. Depreciation of Software (Outdated)			
16. System Test			
17. Inspect Programs			
		Subtotal	

Cost of Nonconformance	Cost per Occurrence	Number of Occurrences	Extended Cost
1. Keeping Track of System Failures			
2. Going Back to Customer to Re-evaluate			
3. Customer Change Requirement			
4. Recode, Debug, Retest			
5. Document Changes			
		Subtotal	

Human Resources
Cost of Quality Categories — Elements

Cost of Conformance	Cost per Occurrence	Number of Occurrences	Extended Cost
1. Prescreen Applications			
2. Interviewing			
3. Personnel Testing (Physical and BOQ)			
4. Reference Checking			
5. Security Clearance, If Necessary			
6. Orientation			
7. Training			
8. Job Description and Work Plans			
9. Safety Program			
10. Quality Improvement Program			
11. Exit Interviews			
12. Performance Appraisals			
13. Attendance Tracking			
14. Productivity Rates			
15. Personnel Record Audits			
16. Tracking of Injuries			
		Subtotal	

Cost of Nonconformance	Cost per Occurrence	Number of Occurrences	Extended Cost
1. Turnover Rates			
2. Grievance Tracking			
3. Nontimely Filling of Position			
4. Cost of Injuries			
		Subtotal	

Activity Based Costing

Information Systems
Cost of Quality Categories — Elements

Cost of Conformance	Cost per Occurrence	Number of Occurrences	Extended Cost
1. Job Descr. (Written)			
2. Hiring and Testing			
3. Schools			
4. Program Documentation and Testing			
5. Cost Benefit Analysis			
6. Risk Analysis of Project			
7. Proper Communication of User Requirements between User and Information System			
8. Verification of Input Data			
9. Test Techniques			
10. Pilots			
11. Parallel Runs			
12. Post Implementation Audit			
		Subtotal	

Cost of Nonconformance	Cost per Occurrence	Number of Occurrences	Extended Cost
1. Systems Do Not Meet User Requirements — Redo			
2. Corrective Maint.			
3. Reruns			
4. Input Cycles Edit and Update			
5. Hardware Downtime			
6. Scheduling Failures			
		Subtotal	

Law Department
Cost of Quality Categories — Elements

Cost of Conformance	Cost per Occurrence	Number of Occurrences	Extended Cost
1. Maint. of Law Library	_____	_____	_____
2. Seminars on Prevention of Product Liability Claims	_____	_____	_____
3. Label Copy Evaluation	_____	_____	_____
4. Adv. Copy Review	_____	_____	_____
5. Safety Program Audit	_____	_____	_____
6. Equal Opportunity Program Audit	_____	_____	_____
7. SEC Compliance Audit	_____	_____	_____
8. Contract Review	_____	_____	_____
9. Checking Paperwork for Errors	_____	_____	_____
10. EPA Compliance Audit	_____	_____	_____
11. Review of Federal/State Submissions (New Products, Patents, etc.)	_____	_____	_____
		Subtotal	_____

Cost of Nonconformance	Cost per Occurrence	Number of Occurrences	Extended Cost
1. Product Liability Matters (Travel, Litigation, Outside Firms, Time)	_____	_____	_____
2. Warranty Reviews	_____	_____	_____
3. Penalties for Late Filing	_____	_____	_____
4. Product Complaint Reviews (Internal and with Regulatory Agency)	_____	_____	_____
5. Product Recalls	_____	_____	_____
6. Defense of Patent Infringement Suit	_____	_____	_____
7. Representing Grievances	_____	_____	_____
8. Internal Dept. Rework (Rewrite, Retype, etc.)	_____	_____	_____
9. Seminars on Defending Product Liability Suits	_____	_____	_____
10. Settlements	_____	_____	_____
		Subtotal	_____

Activity Based Costing

Cost of Quality Totals

	Cost per Occurrence	Number of Occurrences	Extended Cost
Sales and Marketing			
Cost of Conformance	_____	_____	_____
Cost of Nonconformance	_____	_____	_____
Engineering			
Cost of Conformance	_____	_____	_____
Cost of Nonconformance	_____	_____	_____
Industrial Engineering			
Cost of Conformance	_____	_____	_____
Cost of Nonconformance	_____	_____	_____
Procurement and Acquisition			
Cost of Conformance	_____	_____	_____
Cost of Nonconformance	_____	_____	_____
Plant Administration			
Cost of Conformance	_____	_____	_____
Cost of Nonconformance	_____	_____	_____
Quality Control			
Cost of Conformance	_____	_____	_____
Cost of Nonconformance	_____	_____	_____

Manufacturing
Cost of Conformance   _____ _____ _____
Cost of Nonconformance   _____ _____ _____

Comptroller
Cost of Conformance   _____ _____ _____
Cost of Nonconformance   _____ _____ _____

Software
Cost of Conformance   _____ _____ _____
Cost of Nonconformance   _____ _____ _____

Human Resources
Cost of Conformance   _____ _____ _____
Cost of Nonconformance   _____ _____ _____

Information Systems
Cost of Conformance   _____ _____ _____
Cost of Nonconformance   _____ _____ _____

Law Department
Cost of Conformance   _____ _____ _____
Cost of Nonconformance   _____ _____ _____

Total Cost of Quality
Cost of Conformance   _____ _____ _____
Cost of Nonconformance   _____ _____ _____

CUSTOMER SATISFACTION — It is commonly believed that customer satisfaction cannot be measured qualitatively. The assumption is that companies need to rely on their perceptions, rather than facts. We believe that customer satisfaction is measurable and that we can rely on dollar measurements instead of "feel-good" guesses. The way to do this is to make customer satisfaction a comparative measure. In other words, we need to identify what we are spending to achieve satisfaction versus the identifiable costs of customers not being satisfied. In our opinion, *companies who say they don't know what they are spending to satisfy a customer are probably not spending anything at all.* At the same time, they are probably letting the costs of not satisfying a customer go unrecorded as well.

Providing the Information to Control Internal Operations

Day-to-day operations depend on information which is used to plan activities which have the end result of satisfying customers. Too often, however, this information supports plans which are used to direct, instead of control day-to-day operations. Internal operational plans lack an essential type of information which allows us to measure and see how well we are accomplishing specific goals. Current information uses information about labor and material to measure how well we come up to standards. Consequently, operations personnel feel that they need pay attention only to these measures.

Unfortunately, these measures, as we have seen, tend to distort the real picture as they focus on information of little use to management decision-making. Don't be misled by what we are saying. These factors of cost are important, but they are not the only costs which a company incurs. Operations need to be measured for items like total time, quality and their associated costs. Companies that are agile depend on these measures to ensure customer and organizational satisfaction.

Time, for example, should be measured as cycle time — the time it takes to cycle through the production of a part or to complete a service. A standard can then be adopted and the actual time measured against it. In doing this, we must be sure to identify all delays and stoppages as interruptions to meeting cycle time. These interruptions should be identified as nonconformance costs and should be challenged so that your company can start the process of continuous improvement whereby you either reduce or eliminate them.

At the same time, operations should be collecting data which allows for the calculation of dollars associated with both productive and nonproductive time. Some portion of all operations results in nonproductive time and dollars, but the emphasis in total costing is not on highlighting operator or employee performance. Instead, we need to look at what caused the delays and stoppages and to stop blaming the factory worker. Workers, in fact, will be expected to contribute to the improvement effort by helping to locate time and quality cost drivers which can be reduced or eliminated.

Quality is the other internal operation which is on the mind and in the strategic plans of companies worldwide. But, just wanting to make quality parts is not enough. We need to focus our attention at the level of operations by giving workers the opportunity to improve upon their quality performance. Quality goals must be part of the incentive package which is presented to operations. Workers need to understand that for every bad part they produce, one less good part can be made. Thus, quality performance is synonymous with their ability to perform and the ability of manufacturing to be flexible.

The dollars involved in the control of internal operations should be measured for every activity, task and decision made by a company or an operator. Dollars are extremely important to the overall performance of any operation. Above all, they should reflect the ability of an operation to perform within certain expectations or up to certain levels. This performance to cost should not only be the territory of management. It must be expanded to the operator level so that they can influence the cost factors under their control. Instead of revealing cost results to workers, we have emphasized the measurement of efficiency. As we have seen over the years, an operator can always create results which which will make their efficiency look good. But this is at the expense of true cost improvement. Again, it is not the operator's fault. It is the fault of a bad cost system. We need a system which challenges employees to look for improvements. We need cost information which gets collected and reported at the level where it makes the most sense. We need a system that supports company objectives and gets the entire work force aware and contributing.

Expanding the Awareness
of the Elements of Product Cost

Product cost elements have traditionally supported material, labor and overhead measurements. Although material and labor are readily identifiable components of product cost, overhead represents a "catchall" category for all the other costs. To be truly successful, cost systems must break down the components of overhead into elements of product cost that are meaningful at the operational level. Meaningful is the key word here. By "meaningful," we mean that the elements of cost must represent activities which are actually being performed and which therefore represent opportunities for improvement. To use an earlier example, we should be measuring the cost of quality in dollars for each product rather than apportioning inspection and review costs according to an overhead allocation rule. Once again, the intent is to open up the curtains on a window of opportunity.

The cost elements of products or services being performed must also be used by companies to establish rules within an organization. When we were kids, our parents would not allow us to do certain activities based on their cost. We may not have liked it, but there was no choice and we learned to do without. Somewhat the same type of parenting has to be done by companies. But company managers and executives cannot know when to say "no," if they do not have the information they need to make logical decisions. That information comes from measuring the cost elements of a product or service.

As the list of conformance and nonconformance costs at the end of this chapter shows, product cost elements need to be expanded to include factors which affect performance in both negative and positive ways. Only in this manner can we use the cost elements to open our eyes to improvement opportunities. Let's look at a company operation and see how to determine product cost elements and how to control them.

Operation 300 is an assembly work center for motors. Materials are delivered to the operation site where an operator counts, inspects and assembles parts. After assembly, the product is tested again and good product is stacked for delivery to the next work center. Product which fails this inspection are rejected at the component level and then reworked at the assembly level.

What cost elements should be controlled at this operation? The answer is logical if we simply trace what activities affect this operation as the following list shows:

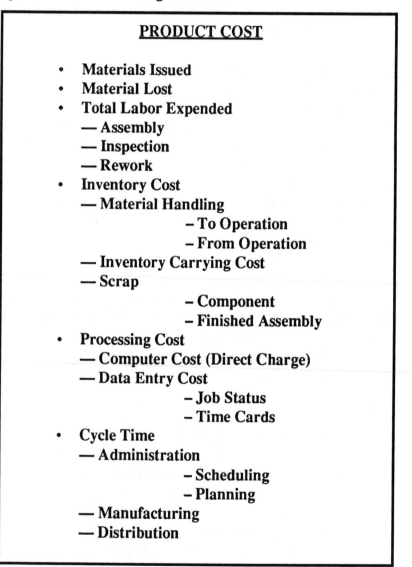

PRODUCT COST

- **Materials Issued**
- **Material Lost**
- **Total Labor Expended**
  - **Assembly**
  - **Inspection**
  - **Rework**
- **Inventory Cost**
  - **Material Handling**
    - **To Operation**
    - **From Operation**
  - **Inventory Carrying Cost**
  - **Scrap**
    - **Component**
    - **Finished Assembly**
- **Processing Cost**
  - **Computer Cost (Direct Charge)**
  - **Data Entry Cost**
    - **Job Status**
    - **Time Cards**
- **Cycle Time**
  - **Administration**
    - **Scheduling**
    - **Planning**
  - **Manufacturing**
  - **Distribution**

Awareness of these elements help employees seek methods to eliminate or reduce cost elements. Workers who are tracking these costs can actually watch them add up and see which activities are contributing the most costs. The more employees in your company who understand this process of identifying the elements of cost, the better able they will be to reduce or eliminate them.

Examining Cost Elements
to Determine Value Added or Lost

When cost elements are identified, they offer us an opportunity to reduce those factors which are not adding value to the product. That is why it is necessary to issue cost reports which reflect these opportunities by categorizing elements into value added or non-value added. Some of our clients do an excellent job at reporting cost in this manner. J.I. Case's internal cost reports break down tractor cost into categories which present a clear picture as to what adds value to the product and what doesn't. As the report on Pages 50 and 51 shows, this method puts the emphasis on reducing or eliminating nonvalue-added costs.

Application
of Direct Traceability of Costs

Cost reduction can only occur when a cost can be directly linked to an activity. Furthermore, it is everybody's responsibility to trace these costs to the activity throughout the organization. For cost reduction to be truly effective, everyone must be involved. The chart on Page 52 shows just how the different activities in a company are related to product cost.

In order to create an organization in which cost is directly traceable to its activity, we need to redefine cost systems. This will

PERFORMANCE MEASUREMENTS
in a
WORLD CLASS MANUFACTURING OPERATION

COST OF PRODUCTION FORMAT IN A WCM COMPANY

I. MATERIAL COST

	Current Period	Prior Period	Changes	Budget	Actual	Better/ Worse
Supplier purchases						
Interplant Requests						
CDC Engines						
Freight						
Exchange						
Subcontract						
Obsolescence						
Inventory Adjustment						
Cash Discount/ Scrap						
Receipts						
Total Material						

II. CONVERSION COST

	Current Period	Prior Period	Changes	Budget	Actual	Better/ Worse
A. Value Added						
Production-Hourly Labor/Benefits						
Salary Wages/ Benefits						
Training Expenses						
Consumed Production Supplies						
Total Value Added						

PERFORMANCE MEASUREMENTS
in a
WORLD CLASS MANUFACTURING OPERATION

COST OF PRODUCTION FORMAT IN A WCM COMPANY

II. CONVERSION COST cont.

	Current Period	Prior Period	Changes	Budget	Actual	Better/ Worse
B. Nonvalue-Added						
Hourly Labor						
Set-Up Time						
Downtime						
Material Handling						
Rework						
Other						
Benefits						
Salary Wages						
/Benefits						
Scrap						
Outside Rework						
Other Gen'l Expenses						
Total Nonvalue-Added	——	——	——	——	——	——
C. Other Costs						
Fuels/Utilities						
Depreciation						
Taxes						
Insurance						
Expense						
Reallocations						
Extraordinary						
Total Other	——	——	——	——	——	——
TOTAL COST of Production	——	——	——	——	——	——

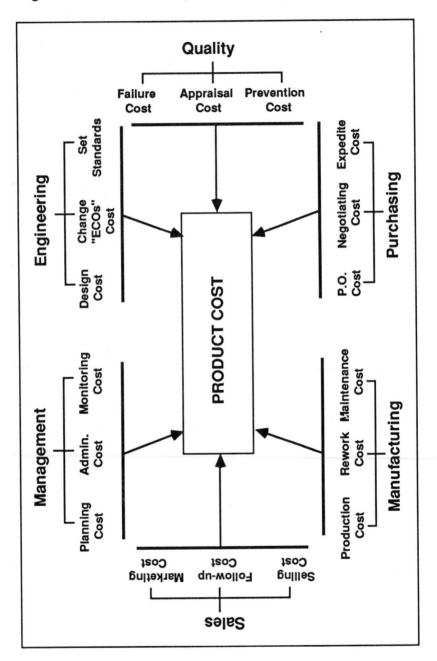

require more direct costing and more intelligent allocations. Functions like Sales, Engineering, Quality and Purchasing which are not used to accounting for their activities will now have to identify how the costs of their activities contribute to product cost. This identification will not require that individuals in these functions start filling out time cards, but attempts must be made to identify tasks performed in the day-to-day operations of their jobs. And then, most importantly, we need to start associating costs to these tasks and then assigning the costs directly to the lowest practical management reporting level.

Let's look at preparing a purchase order for an example. The following list identifies the tasks:

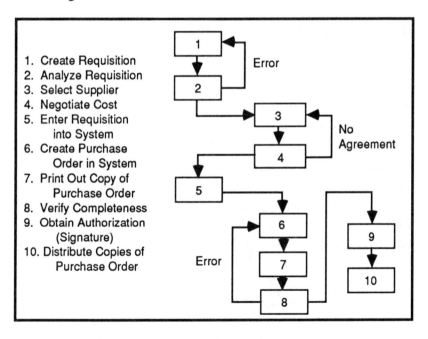

1. Create Requisition
2. Analyze Requisition
3. Select Supplier
4. Negotiate Cost
5. Enter Requisition into System
6. Create Purchase Order in System
7. Print Out Copy of Purchase Order
8. Verify Completeness
9. Obtain Authorization (Signature)
10. Distribute Copies of Purchase Order

All of these tasks represent cost-expending activities whether they are associated with labor, paperwork, computer processing or mailing. Gathering information on the costs of these activities allows us to benchmark the cost of preparing a typical purchase order. Once that is established, we can then calculate the total cost by multiplying the benchmark by the number of purchase orders prepared in the building of a certain product. In that way, we can trace this product cost back to a purchasing activity and seek to reduce or eliminate it. The method described may not be perfect, but it is better than ignoring the cost of tasks altogether. If used properly, the new method will draw attention to the cost of activities and then the information which is gathered can be shared with people who can find ways to eliminate the task or reduce the costs associated with the task.

Implementation of Cost Reduction Strategies

The first requirement you must meet in the implementation of a cost reduction strategy is to be serious about your efforts and to realize that none of them will result in a quick fix. One CEO was recently overheard to say that for each $100,000 the company saved, it resulted in an increase of $0.01 to the company's earnings per share. Next, you will need to realize that traditional strategies of reducing costs by attacking the cost of labor and material have been actually proven to generate more costs than savings. As you now know, that is because the real opportunities for cost reduction are behind the scenes.

After tracing costs back to the activities which are responsible, we need to evaluate costs so as to determine which opportunity to reduce first. At that point, we can then begin fashioning goals and a plan of attack for a controlled reduction so that we know what is working and why. Don't forget that some activities and their associated costs don't need reduction. They can often be completely eliminated with no detrimental effects since they were never required to begin with. In fact, these opportunities which we call "waste" should head the list in our strategy to reduce. Other activities will require more thought in devising a cost reduction plan which achieves the desired savings without damaging results or quality.

Any plan we develop must challenge the workers to reduce the elements of cost which erode our profit margins. It will not only require teamwork, it will prosper with employee involvement. By creating a system which accurately captures, allocates, sorts and distributes cost results by area of responsibility, we can build an environment where people are encouraged to take up the challenge. Part of this total involvement program will involve setting goals of expected improvement in cost reductions. These goals should be realistic and obtainable. If people feel that a goal is not realistic, little or no effort will be made to seek improvement.

Chapter Three

IDENTIFYING
THE COMPONENTS
OF TOTAL COST

Trying to run a company under a traditional cost system is something like driving a car on a dark night with all the windows fogged up. You can make out some shapes ahead of you if you wipe wet circles on the windshield with your sleeve, but not very clearly. Likewise, the traditional arrangement of cost categories into material, labor and the amorphous category of overhead doesn't begin to give a company the visibility it needs to run a cost-effective and competitive business in today's marketplace.

Activity Based Costing (ABC), on the other hand, is like driving the same car with the defroster going full blast. Now, the cars and trees and pedestrians ahead are more than dark shapes or blurs. Now, you have a clear picture of where you are going. Our point is this: The more details a cost system collects and reports, the more visibility an organization has into the process of making decisions. Business decisions, such as determining what products are profitable, what products should be discontinued, what assets are effective and what assets need to be retired, are all too commonly made without correct and timely information. And our strong contention is that management and accounting cannot be sure if the organization is doing all that it can to be competitive, if it is driving along without the correct information.

As a first step, a total cost system like we are proposing would break overhead down into a number of cost categories (see Figure 3-1) which more accurately reflect actual contributions to total cost.

We will be discussing each of these areas in this chapter, but let's first look at how all this detailed information will be collected. Most managers and executives probably think that they need to hire more people, buy more computers and software and invest in state-of-the-art operational systems. It is no surprise then that we so often hear that the costs of collecting more data far outweighs the benefits which can be derived from more reported results. Peter Grieco has always said that too many companies are data rich and information poor. In a poorly designed operation, this may indeed be true. Such an inefficient company would require people to report more information by adding on to their already heavy work load. But this is not the only way to obtain valuable accounting information.

CATEGORIES OF COST

Cost of Quality

Appraisal Cost
Failure Cost
Prevention Cost

Cost of Inventory

Carrying Cost
Handling Cost

Cost of Design

Planning Cost
Documentation Cost

Cost of Selling

Advertising
Selling Cost
Customer Service

Cost of Engineering

Redesign Cost

Cost of Administration

Supervision
Facilities
Utilities

Figure 3-1

In a World Class company, people in the organization work toward making products of high quality in an environment where costs are being continuously reduced. In order to accomplish this task, they have not gone out and bought ideas and techniques, they have forced themselves to be smarter, more creative and to strive for simplicity. This is what a modern cost management system must do as well. It needs to find simple, but powerful ways to collect the already existent data in our operations.

The first obstacle to be overcome in this quest for better ways of collecting data is the stigma that accounting traditionally places on creative methods and results. There is the belief that new ways of looking at data are not showing real information or that the data are reflecting what we want to see rather than what is actually going on in the business. Activity Based Costing is not trying to fabricate results to make a company look good or bad. It is looking to produce results which are more reflective of and relevant to actual performance. It is in this sense that we say ABC results are more creative because now people have information which they can use in creative ways to identity and then reduce the components of cost. The traditional method of allocations and statistics, as we have seen, serves only to hide true costs.

In this chapter, we will look at the components of total cost and show that there is a real need to collect more detailed data and use it in cost calculations which generate more useful output. This output can then be used in creative ways to help a company in its world class improvement efforts. Such company activities as quality, inventory, purchasing, compliance and production are highlighted as areas which can be improved by collecting more effective cost data.

Cost of Quality — Understanding the Value of Quality Improvement

The Cost of Quality (COQ) is a subject of wide interest as is evidenced by the large amount of material which has been written on the topic and by the large number of companies becoming involved in Total Quality Management (TQM) programs. There are also a number of well-publicized quality prizes, such as the Deming and Baldrige awards, which keep interest in quality high. And lastly, there is the ISO 9000 which has established quality systems criteria which companies must meet in order to trade. All of this activity forms a good foundation upon which companies can initiate efforts to obtain ABC results. This is the way that many of our clients started to address total cost issues.

Unfortunately, many Cost Managers do not understand Cost of Quality and how important it is to organizational results. COQ is not just the capturing of nonconformance data such as scrap, unfavorable yields and rework. Organizations should also use COQ to measure the relationship between dollars spent due to the lack of quality and dollars spent on prevention or corrective cost in solving quality issues. Let's look at a COQ report in order to see what we mean by analyzing organizational results.

When you start evaluating the reported results in Figure 3-2, several issues become apparent:

1) The company has not managed to control its internal failure costs. Note how the dollar amounts are increasing.

2) Although external failure and appraisal costs are remaining about the same, neither seems to be improving.

3) Prevention costs are the lowest costs, which suggests that the company is not spending any

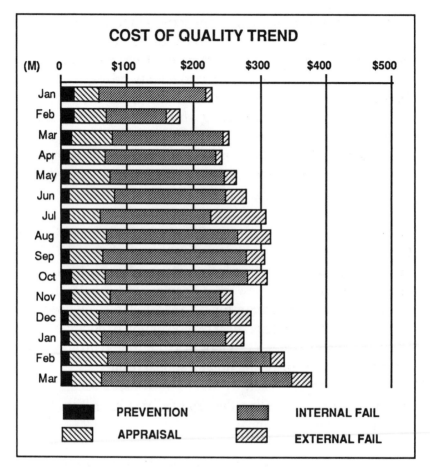

Figure 3-2

more dollars than they must. More dollars spent in the prevention area would translate into improvements in the other three categories.

4) The company can probably expect the COQ to continue to rise until more dollars are devoted to improving quality.

Now that you have an idea of how the different elements of the Cost of Quality relate to each other, let's take a closer look at these components themselves. First, most experts use four areas to classify quality costs as shown in Figure 3-3.

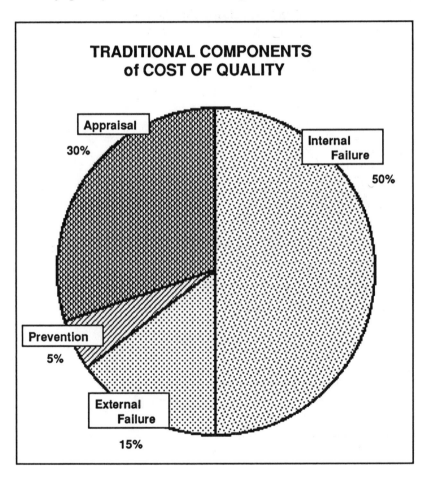

Figure 3.3

We believe that this traditional method of capturing and analyzing the elements of quality cost should be expanded to five areas in order to better reflect what is driving costs. Again, this is a situation in which the more we know, the better able we are to reduce costs.

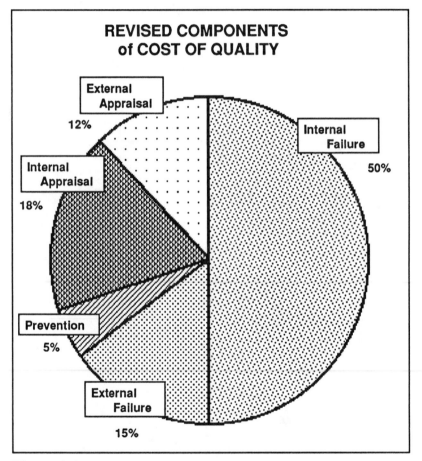

Figure 3.4

Now let's look at each component in turn. It is believed that the largest component of COQ in most companies is **failure costs** which, when collected, are broken down into **internal** and **external costs**. **Internal failure costs** typically account for most of a company's nonconformance issues. We recommend that you use a great deal of care in collecting internal failure data. The classification often becomes a black box into which companies dump any unexplainable cost. ABC's goal is to do just the opposite and have no unexplained costs. Critics of ABC contend that this is no more than moving money from one bucket to another, but this is not true. ABC does much more than move dollar amounts. It **illuminates** areas of opportunity. It is not simply bean counting, but the exposing of cost drivers which can then be either justified, eliminated or reduced. That is why it is imperative that you make sure that all costs which you collect are traceable to the activities being performed.

You also need to be as objective and realistic as possible in spelling out the components of internal failure costs. Some realistic costs are:

> **Scrap** — The amount of unplanned material which is not recovered and therefore disposed of. It is typically accounted for through some percentage which is used to cost out the product. Some organizations call this cost salvage.

> **Rework** — The dollars associated with correcting nonconforming product.

> **Warranty** — The cost associated with items which are returned by a customer for correction of nonconforming product. This cost should include the expense to make the correction, replacement cost and freight.

> **Machine Downtime** — The cost associated with

> unplanned or unexpected downtime. This cost
> should include the emergency maintenance
> cost and the loss of product cost.

Remember to capture the material, labor and burden costs for each of these categories as well.

External failure costs should be related to costs which can be directly traced back to a supplier's nonconformance. For example, the cost of any materials received which are determined to be defective should accumulate in this category. External failure costs should also include the dollars related to returning material, line failures and products returned by customers.

Appraisal costs are usually the second largest component of COQ. Simply put, appraisal costs relate to any expense associated with inspection or testing to ensure conformance. A total cost system, of course, tries to convince companies to spend less money in this area and more money on preventing defects. In addition, Activity Based Costing suggests that this category should be separated as well into internal and external components. When this is done, companies typically look at the amount of dollars spent on inspecting suppliers' materials and begin to increase their efforts to develop stronger relationships with suppliers through Supplier Certification programs.

Companies have made little effort to improve COQ by spending more on **prevention costs** for the reduction or elimination of nonconformance. This category is also viewed as a quick way for management to save money by cutting out some of the activities associated with prevention. This shortsighted mindset identifies these costs as nonvalue-added when, in fact, prevention costs can be demonstrated as saving dollars in the area of failure costs.

Instead, management needs to refocus their mindset so that more and more dollars are put into such popular and beneficial activities as improvement teams, internal and external training programs, supplier certification programs and zero-defect programs. Thus, even though prevention costs may rise, total cost of quality will go down.

Cost of Inventory — Developing
Cost Factors as an Incentive to Improve

Organizations measure inventory in a number of ways — inventory dollars, inventory turns, obsolescence dollars, slow moving dollars, and days, weeks or months of inventory. Although these measurements arrive at figures, they are not accurate or thorough enough for a total cost management system because none of them capture the cost of carrying inventory. For instance, some accounting systems use a 2% a month or 24% a year figure as a default to the cost of inventory (COI). It has been our experience, however, that most of our clients's COI is higher and that the drivers of inventory costs need to be identified. In fact, we have consistently noticed that the COI at our clients ranges between 25% and 45%. The numbers tell the story. A measurement for the carrying cost of inventory must be calculated and analyzed by ABC reports, such as the sample report at the end of this section.

As the sample report which follows on page 70 demonstrates, the categories of COI should include:

- **Storage Space.**

- **Handling and Equipment.**

- **Inventory Risk.**

- **Taxes and Service.**

- **Capital Cost.**

Storage Space Cost — In order to house inventory, you need space and space costs money. In an ABC system, these costs are directly linked to the units that occupy the space. Therefore, the following data is needed to determine storage costs: Taxes, depreciation, maintenance and repairs, utilities, janitorial services and security. In addition, all locations where inventory is stored must be considered and should include internal and external cost elements such as the following: Warehouse, stockroom, yard space, trailers, quality areas, material review area, receiving and shipping.

Handling and Equipment Cost — This cost area should include all equipment used to move inventory in and out of storage plus the equipment's depreciation, fuel, maintenance and repair, insurance and taxes. In addition, equipment used to test, weigh, wrap, open or print (bar codes) should be recognized.

Inventory Risk Cost — Whenever inventory exists, so do expenses associated with risk. And the more inventory, the higher the cost. Inventory insurance, for example, is based on inventory value. Pilferage is another risk cost which can be either a relatively minor cost or a major one. One of our clients, for instance, discovered that water treatment equipment worth $3.5 million had disappeared from their building. Inventory which is kept in stock for long periods of time also runs the risk of becoming obsolete or deteriorated physically to the point where it cannot be used. For some reason which we don't quite understand, companies love to keep obsolete inventory. (We really know why — it's an asset.) We are constantly trying to change their mindset on this hoarding by pointing out how much it costs them to hold on to something that they will never use. One final risk which must be taken into consideration is the losses which result from inventory pricing going up and down.

Taxes and Service — COI also includes the taxes which must be paid on inventory and the labor costs for handling the material. Don't forget to pick up the costs of benefits and fringes in the labor cost as well. Another expense in this COI area is clerical costs which are often not included even though they can be quite large. In fact, in an ABC system, clerical costs are much larger because they include all associated activities, such as entering transactions, producing forms and bar codes, researching purchase orders, calling truckers and scheduling deliveries.

Capital Cost — In calculating capital cost, the cost of money can be used as either a dollar figure or as a percent. The area should also include the interest paid on money borrowed to invest in inventory, equipment, land, buildings and storage areas.

Activity Based Costing

SPECIAL SECTION

How to Calculate
the Costs of Carrying Inventory

The following section is divided into two parts. The first part is a form for recording the amounts which you obtain. The second part is a line-by-line description of how to fill in the first part.

I.   STORAGE SPACE COSTS

DOLLARS

1. Taxes on land and buildings for stores _____
2. Insurance on storage building _____
3. Depreciation on storage building _____
4. Depreciation on other warehouse installations _____
5. Maintenance and repairs of other buildings _____
6. Utility costs, including heat, light and water _____
7. Janitor, watchman and maintenance salaries _____
8. Storage/Handling at other locations _____
   Subtotal: Storage Space _____

II. HANDLING EQUIPMENT COSTS FOR STORES
     ONLY (not including central trucking)

DOLLARS

9. Depreciation on equipment _____
10. Fuel for equipment _____
11. Maintenance and repair of equipment _____
12. Insurance and taxes on equipment _____
   Subtotal: Handling Equipment _____

III. INVENTORY RISK COSTS

DOLLARS

13. Insurance on the inventory _____
14. Obsolescence of inventory _____
15. Physical deterioration of inventory, incl. scrap _____
16. Pilferage _____
17. Losses resulting from inventory price declines _____
   Subtotal: Inventory Risk _____

IV. TAXES AND SERVICES COSTS

DOLLARS

18. Taxes on inventory _____
19. Labor costs of handling and maintaining stock_____
20. Clerical costs of keeping records _____
21. Employer contributions to Social Security
    for all space, handling and inventory
    service personnel _____
22. Unemployment Compensation Insurance for all
    of the above personnel _____
23. Employer contributions to pension plans, group
    life, health and accident insurance programs
    for above personnel _____
24. A proportionate share of general administration
    overhead, including all taxes, social security,
    pension and employer contributions to insurance
    programs for administrative personnel _____
    Subtotal: Taxes and Services _____

V. CAPITAL COSTS

DOLLARS

25. Cost of money _____
26. Interest on money invested in inventory handling
    and control equipment _____
27. Interest on money invested in land and buildings
    to store inventory, if owned _____
    Subtotal: Capital Costs _____

GRAND TOTAL: _____

28. Average inventory on hand for storerooms
    considered in the analysis above (in dollars) _____
29. Calculate carrying charge percent by dividing #28
    into Grand Total Amount _____%
30. Current cost of money (in percentage) _____%
31. For **TOTAL CHARGE** (in percentage), add #29
    to #30 _____%

Activity Based Costing

TOTALS

	DOLLARS
I. STORAGE SPACE COSTS	_____
II. HANDLING EQUIPMENT COSTS FOR STORES ONLY	_____
III. INVENTORY RISK COSTS	_____
IV. TAXES AND SERVICES COSTS	_____
V. CAPITAL COSTS	_____
TOTAL COSTS	_____

DEVELOPMENT OF INVENTORY CARRYING COSTS

Line No.

1. Estimate of share of real estate taxes paid for the part of the building and land occupied by storage facilities, including inside and outside.

2. Estimate of share of insurance for same areas. NOTE: many large firms are either self-insuring or carrying a large deductible to cover catastrophes only.

3. Annual depreciation actually claimed for the building/land used for stores.

4. Annual depreciation on any other remote locations or temporary locations used for storage.

5. Estimate of annual maintenance costs spent just for storage areas. (If outside, include some snow removal, if appropriate for your area.)

6. Estimate of annual maintenance costs spent just for the storage area buildings.

7. Estimate of share of annual expenses for security and janitorial for stores area.

8. Annual costs incurred for storage and handling at other location.

9. Annual depreciation on forklifts, cranes, stacker/pickers, racks, etc. for all equipment used for handling inventory in store area. Do not include any equipment from central trucking, shipping, receiving, operations, etc.

10. Annual estimated fuel costs for above.

11. Estimated annual maintenance costs and/or maintenance contracts for items in No. 9 above.

12. Annual insurance and taxes on equipment items in No. 9 above, if known.

13. Annual insurance premiums paid to cover casualty losses of stored inventory (not Work-In-Process), if any.

14. Estimated annual write-offs due to obsolescence. (Note: if written off, material must be destroyed and/or disposed of.)

15. Estimated inventory losses due to scrap generated in stores handling, shelf life deterioration, etc. Do not include shop generated scrap or material scrapped as a result of inspection design change.

16. Estimated inventory losses due to employee pilferage for personal use.

17. Estimated reduction of inventory value because of price decline—use the lower of cost or market value.

18. Taxes on inventory are no longer assessed in many states. But when inventory is stored in another state, it would be checked to see if any entry should be made on this line.

19. Labor cost to store inventory. This is the BIG ONE, and should include the annual wages and salaries of all individu-

als connected with storing inventory. (Do not include material handling such as shipping, receiving, or trucking.)

20. Estimated clerical costs for data entry, cycle counting, reconciliation, error correction, document handling, etc., including time and cost of computer operation, maintenance and reports.

21-24. Fringe benefits. These can be combined and entered into No. 24.

25. Generally thought of as the cost of money. Can be handled in No. 30 as a percentage figure, but not in both places.

26. Estimated cost of interest expense on handling equipment purchases; or, the interest that could have been earned on the money which is tied up in storage handling equipment (related to No. 9 above).

27. Estimated interest cost (either paid or opportunity lost) on the money tied up in storage buildings and land. (Related to No. 3 above.)

28. Since the cost figures developed in the above are annual and apply to the inventory assessment across the year, an average investment-on-hand figure should be determined for the storeroom materials (not WIP or FG).

29. Develop a percentage figure by taking the annual costs (total) over the average investment.

30. Then add the current cost of money, usually thought of as the current prime rate. Sometimes, the actual cost is more or less than the prime, but the prime rate is normally close enough. (DO NOT DUPLICATE line 25 here; it is one or the other.)

31. The total actual current inventory carrying charge is then the sum of line 29 and line 30.

After accumulating all the components of the COI, a carrying cost can be calculated (see lines 28 to 31 on the facing page). Use the average inventory value for the areas being considered. It is not necessary, either, to use all inventory to calculate the carrying cost. You can calculate different areas separately in order to emphasize an inventory area in great need of reducing. In fact, the cost of carrying inventory can be used as a tool to convince people throughout the organization that levels must be reduced. Let's look at an example of just how powerful an argument these figures can make.

A company has accumulated an inventory that is valued at $10 million. The inventory carrying cost was then calculated to be 30%. The cost of carrying the inventory would thus be $3 million a year.

$10,000,000 X 30% = $3,000,000

To make this number even more dramatic, we recommend that you then calculate a daily rate which, in this case, is .0008% a day. Thus, the **daily** cost of carrying inventory turns out to be:

$8,427 a day.

With dollar amounts like this, how can we not be looking to reduce inventory?

In summation, we believe that your organization should not only be looking to reduce inventory levels, but also to reduce the components of the inventory carrying cost. We encourage all of our clients to examine these components and find ways to reduce. One of our clients was able to reduce costs in a very simple but effective way. This company was using 17 forklifts to move materials from the receiving area to a process line at the opposite end of the facility. After looking at the problem with us, the company decided to construct a new loading dock closer to the process line. The net result was that the company now needed only one forklift and substantially reduced its carrying cost for equipment.

Cost of Procurement — Focusing on the Expense of Acquiring Materials

The challenge facing us is to procure components, raw material or services for our plants to satisfy demand. All too often, however, we meet this challenge with a traditional mindset. The traditional practices of supply management would have us believe that:

> - **It costs less to buy more.**
> - **Price is more important than quality.**
> - **More frequent deliveries cost more.**
> - **Suppliers are adversaries.**

In order to change this mindset, the primary goal of Purchasing should be to implement supply management programs which develop long-term relationships or strategic partnerships (but not necessarily in the legal sense) with major suppliers. These programs should not only look at improving cost, delivery, quality and quantity, but at improving our buying practices as well.

In our book, **Supplier Certification II,** *A Handbook for Achieving Excellence through Continuous Improvement,* Peter Grieco shows in detail how to implement a program to develop close relationships with suppliers which are based on trust and communications. One of the methods advocated is an agreement in which both parties share the risks as well as the profits. In this new method of buying, cost accountants need to develop partnerships which identify both the supplier's and the customer's concerns. The supplier and buyer must negotiate shared responsibilities and work toward a win/win situation.

For instance, it would be beneficial to both if quantity could be predicted over a defined period (one year, six months, or a quarter) by spreading the quantity over that period using a predicted forecast. The contract price would then be established for a period quantity. Obviously, this method is of great value to the supplier, but the buyer is safeguarded as well by not being locked into 100% of the quantity.

Furthermore, both parties would agree to the following:

- **Suppliers should not build ahead.**

- **Changes in quantity can only be made within an agreed ratio.**

- **Buyers must communicate timely buys within the negotiated lead time.**

- **Purchases of more than required quantity must be negotiated separately.**

This type of agreement should allow both parties to work together in confidence as the agreement shows on the next page.

CONTRACT BUYING

Contract Quantity   6,000 pieces
Contract Period   6 months
Lead Time   60 days

Month 1	Month 2	Month 3	Month 4	Month 5	Month 6
1000	1000	2000	1000	500	500

Agreed Ratio

Days	Ratio
0-30	100%
31-60	80%
61-90	50%
91-120	20%
121-over	0%

Figure 3-5

What this contract ratio means is that the buyer must commit to 1000 pieces during the first 30 days or 100%. In the next 30 days, the buyer can adjust the quantity by 20%. In other words, we can buy as low as 1800 pieces without violating the contract. During the third month, the buyer can adjust the quantity by 50%, by 80% in the fourth month and can cancel all quantities in the fifth and sixth months. As you can see, this arrangement reflects a total commitment of only 3000 pieces during the contract period, but the buyer can purchase up to 6000 pieces under the same terms. If the buyer should require more than the quantity agreed upon in the contract ratios, then a provision should be in place to change prices either plus or minus for the order. If the buyer's demands for quantity increase or decrease over the contract period, it should be their responsibility to communicate changes within a specific period of time.

In this type of contract, neither partner is expected to store any inventory, thus lowering costs for both sides. Overall, this buying method demonstrates a win/win relationship since it helps the supplier plan production with more accuracy and locks in costs more effectively for the buyer.

There are also value analysis techniques which must be made part of the agreement in order to save additional money. It entails, once again, the collection of more detailed Activity Based Costing reports to identify where opportunities for improvement exist. As we have said before, this will require a shift in the departmental mindset away from cost aggregates. In work with our clients, we have seen that a close look at day-to-day activities reveal a great number of opportunities to extract savings. Consider the following purchasing activities and how they present us with considerable cost targets:

What does it cost?

- Activities engaged in determining demand to authorizing a purchase typically include: analyzing MRP (Manufactuing Resource Planning) reports, preparing requisition forms, distribution of paperwork to Purchasing, validation of request, and authorization to buy.

- Preparing a purchase order through the mailing

and distribution of copies typically includes: sourcing suppliers, negotiating price and terms, entering data into computer, correcting entries, printing the purchase order, separating copies, mailing copies to supplier, distributing copies within company and filing forms.

• Purchase order follow-up (expediting) typically includes: analyzing reports through determining commitment.

The demonstrations above make it obvious that the same types of expenses are incurred with all the activities which are performed. This is not just a coincidence. When a company starts to produce more Activity Based results, it will become evident that the same components of cost are captured throughout the organization:

- **Time.**
- **Materials.**
- **Labor.**
- **Utilities.**
- **Waste.**

Cost of Processing — Revealing the Cost of Data Transactions

For those who believe that the cost of computers is coming down, we recommend that they look at the cost of processing of data within their organizations. The fact is that we are depending more and more on computer reports to run our businesses. That, in itself, is not the problem. But the amount of time and effort being expended to capture and input the data needed to generate these reports is an issue that needs to be addressed. Take a moment to think about the number of transactions being processed by your computers each month — 5,000, maybe 10,000 and perhaps even many more. Now take another moment to consider how these transactions were created. In most organizations, they are the result of direct input into the computer system by people performing activities within the company.

Our concern is not over the need for data, but over how it is collected. Too often, the data is collected and input by people who are responsible for production and not for data processing. Activity Based Costing looks to cost out the activities connected with providing data to the computer in the hopes of encouraging companies to seek better ways of capturing the required data.

"Wait a minute!" you are probably saying to yourself at this point. "Didn't you just say that an ABC system wants more detailed data? Now you are telling me that it costs too much to collect. How can my company do both?"

There are many ways to achieve the goal of satisfying the need for more data while cutting costs. But first, you need to realize that the real enemy of computer processing cost is people time. In other words, we want the information; we just don't want people to input data. Let's look at why. The following examples use activities to demonstrate how organizations are losing productive time:

- Entering production labor on time cards and then manually inputting this data into a computer.

- Entering production results such as completions and then manually inputting this data into a computer.

- Entering nonconformance such as scrap, rework, downtime, etc.

- Entering material transactions such as receipts, stocking, kitting and physicals.

And then there are activities used in support efforts:

- Preparing bills of material, routings, work orders, schedules, master schedules and maintenance requirements.

And finally there are activities used in administrative efforts:

- Preparing financial entries, invoicing, vouchers, producing checks and entering budgets.

- Creating sales orders, shipping papers, product quotes and cost estimates.

The list above barely scratches the surface of the number of transactions which are being manually input into the typical company's computer system. It is the task of World Class companies to review the need for data and to challenge current methods in order to get people out of the inputting business. With the capability of today's computer and its associated hardware, there is no reason why most companies cannot take advantage of a computer system's ability to input data automatically.

At one of our clients, for example, we measured the time and effort associated with the activities which began with receiving all the way through to distribution to the floor or stockroom. We discovered that half of the time was spent performing activities related to processing data, while the other half was related to handling materials. We convinced our client to implement a bar coding system for many of the nonvalue-added requirements. The result was an increase in productivity, as the time spent on data input shrunk to 1/8 of the total time, that is, from 50% to 12.5%.

Another one of our clients who was concerned about data processing began to record the amount of time spent by factory workers capturing shop floor data such as material usage, labor and completion transactions. The results convinced the company to implement a backflushing system which automatically collected and inputted the above data based on product bills of material and costed routings. The effort showed immediate improvements in productivity.

Yet another client analyzed its purchasing function and the number of hours expended on inputting transactions and printing forms. The client's response was to introduce Kanban cards with bar codes. Together, the two developments eliminated 90% of all transactions and four of the forms which were traditionally used. The Kanban cards at this company served multiple functions:

- **Purchase order.**
- **Packing slip.**
- **Receiver.**
- **Stocking ticket.**

The cost savings on forms processing alone paid for the bar coding equipment at this client's company.

As you can see, costing out the activities we have been discussing takes imagination and the initiative to change your mindset to one that seeks cost-saving opportunities. We tell our clients that the best way to begin this task and arrive at a cost for any activity is to take a clipboard and go out and observe what people do. Record the activity and the time it takes to complete as shown in the example below:

RECEIVING	
Activity	**Time**
1. Get packing slip from delivery	2 min.
2. Visually verify that delivery is ours	1 min.
3. Find copy of P.O. in file (or verify on system)	10 min.
4. Produce receiving report	8 min.
5. Count parts	10 min.
6. Wait for quality to inspect	24 hrs.
7. Distribute to stock (or to shop floor)	20 min.

Figure 3-6

If you believe that time is money, then you can see from our example above that there is a lot of money to be saved. The next step would be to assign a cost to labor and waiting as well as computer processing and materials (forms). This enables a company to arrive at a cost for receiving that can then be applied to every receipt that is processed in the company.

Cost of Administration — Revealing the Cost of Running the Business

In a traditional company, administrative activities are perceived as a burden which is collected as an aggregate figure to be absorbed according to some overhead formula by all the products manufactured and sold. In a company using an activity based costing system, these same administrative costs are distributed far more equitably. An effort is made to collect and report the total cost of each individual product and to assign administrative costs only in ratio to the amount of time the product consumes. In order to effect a proper distribution of administrative cost, two factors must be determined:

• **What cost components should be included in a consideration of administrative costs?**

• **What should the basis be for determining the product to which these costs are assigned?**

Some of the cost components of administration will probably include salaries, materials processing (reports, graphs, financial papers), legal fees and advertising to name only a few. Your cost elements will vary depending on the organization of your company and the activities it performs. The only criterion which is the same for all companies is that all components of costs must exist currently. If cost components are now part of an aggregate figure, your first step is to discuss how to separate the different elements and assign them to the appropriate products or services. In other words, like all activity based costing, your goal is to report costs more directly.

The basis for determining the product to which these administrative costs are assigned depends on the activity that generates the cost. The goal is not to capture every dollar and penny exactly. However, an attempt is made to make the spread more equitable than traditional methods. This means that the basis can vary as long as it logically links to the tasks being measured. This is the advantage of ABC over traditional cost techniques: There are not generally accepted practices in ABC which force companies to make accounts comply with a technique that is neither relevant or practical.

Let's look at effort as an example. It is not collected directly in a traditional costing environment, so its associated costs must be allocated. But let's look more closely at how effort associated with performing administrative activities can be captured so that the activities' costs can be determined. There are several methods:

- **Profit.**
- **Transaction volume.**
- **Sales volume.**
- **Number of orders.**
- **Number of computer terminals.**

Imagine a company which processes on the average 500 vouchers per month. In month seven of last year, the company ran a report writer program which totals all the transactions created by the accounts payable voucher entry program. The report writer keyed in on the accounts payable's ID field and selected all transactions with "VE" (voucher entry). The report writer also keyed in on the product type field in order to determine the total number of voucher entry transactions for each product type. The results are shown in the chart on the next page.

Product	Transactions	Ratio
100	53	8%
200	76	12%
300	393	61%
400	126	19%
	648	100%

The ratios from the table above are considered to represent the effort expended by the Accounts Payable department for each product. The company's total expenses for Accounts Payable during month seven was $8,876. By applying the product ratios calculated above, the company was able to cost out the Accounts Payable total expenses for each product as follows:

Product	Ratio	Cost
A	8%	$ 710.08
B	12%	1065.12
C	61%	5414.36
D	19%	1686.44
	100%	$8876.00

The company's report writer then posted the ratios to a statistics file which used the figures in month-end allocations in order to create new cost categories for total product cost reports. The company in this example used a variable method to calculate Accounts Payable expense distribution since the expenses can and do vary from month to month. Other companies may determine that fixed ratios can be posted in the statistics file when there is very little change in volume from month to month.

As for other administrative costs, other methods need to be utilized as long as they hold to the rule of being relevant, logical and practical. To this, we would also add that the method should allow visibility into the activity so that a continuous improvement can be implemented. Remember that the ultimate goal of an ABC system is to reduce or eliminate any waste associated with a measured activity.

Cost of Compliance — Internal and External Cost of Being On Time

One of the greatest opportunities for a company to pursue improvement is in the area of compliance performance. What exactly do we mean by the cost of compliance? Put simply, this is what we mean:

**The Cost of Compliance
measures the dollars associated
with *not* doing
what is expected.**

Figure 3-7 shows just what some of these activities are.

This list, of course, could go on and on through every function, activity and task in your organization. The focus of compliance is on showing that there is a cost associated with *not* doing what is expected or in the time allowed.

At one of our clients, Industrial Drives, the president, Charles Perry, was becoming concerned that his organization was not

COMPLIANCE ELEMENTS

Meeting Standards

- Time
- Cost

Delivering On Time

- Internal
- External

Production Levels

- Time
- Quantity
- Quality

Administrative Performance

- Goals
- Sales
- Budget

Schedule Attainment

- Time to sell
- Time to develop
- Time to production
- Time through production
- Time to ship

Figure 3-7

meeting schedules. He asked us to find out why his organization always waited until the end of the month to ship product. "I want my organization to do today's work today!" he informed us. We helped set up a team (called "Today's Work Today") whose objective was to discover the reasons behind the issue of not meeting schedules. In fact, what the team was looking at was a compliance issue which started at the back door where product was shipped and traced itself back up front to where orders were being taken in sales.

In most organizations, the area of concern would be at the back end of this production cycle and therefore all of the attention would be directed there. But, the real roots of why the product doesn't go out the back door on schedule reflect every facet of cycle time, as shown in Figure 3-8. Instead of concentrating on one area, organizations should investigate all of the functions and activities throughout the cycle in order to determine compliance. You should be asking questions in particular which seek to determine cost and time. For every noncompliance in a cycle step, production time is cut. In fact, this is such an accepted practice that expediting almost always occurs nowhere else but in materials and manufacturing.

Remember, too, that even though the cost of compliance can be calculated based on time, it can either be converted to dollars or calculated strictly on dollars right from the beginning. For example, a cost can be determined as a dollar value per day or, when appropriate, per hour. The cost of compliance is then calculated by multiplying the dollars by time. To determine the dollar amounts, a company need only look at the expenses associated with having to make up time, whatever method is used for collecting dollars relevant to compliance.

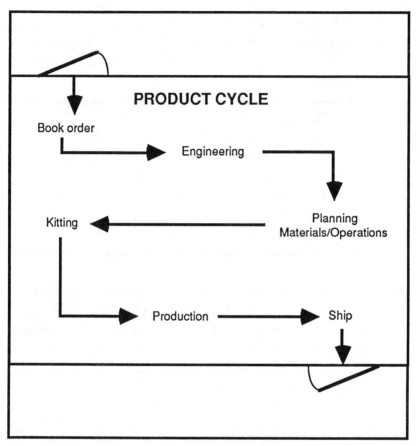

Figure 3-8

Cost of Production — Revealing
the Cost to Get Product out the Door

Now let's focus our attention completely on the costs of production which measures the dollars associated with manufacturing products. The objective of ABC in this area is to collect the *total costs* for all functions and tasks associated with a product. These results are then compared against standards in order to measure financial success more realistically. The actual components of cost include both value and nonvalue-added costs and fall into the following categories:

- **Material cost.**
- **WIP carrying cost.**
- **Direct labor.**
- **Indirect labor.**
- **Set-up.**
- **Downtime.**
- **Maintenance (preventive).**

Material cost — Material cost needs to go beyond the cost of raw material and the cost of purchased parts. ABC seeks to highlight all the costs associated with handling and accounting for material as well. Thus, the ABC list of Material cost would include:

- **Receiving.**
- **Stocking.**
- **Handling.**
- **Picking (kitting).**
- **Outside vendor handling.**
- **Obsolescence.**
- **Returns (to stock).**
- **Inventory adjustments.**

Work-in-process carrying cost — Materials on the floor in queue or waiting to be moved fall under inventory carrying cost and should be associated with the product for which they are being staged.

Direct labor — No mystery here. Most companies already have a method to collect and report labor directly as it relates to operations in the manufacturing area. In service businesses, each labor task must be identified.

Indirect labor — This element of cost is usually perceived as not worth capturing directly to production activities and is usually captured in overhead. As we have already seen, an ABC system does not overlook this area of opportunity and looks at components such as the following:

> • **Managerial support.**
> • **Material handlers.**
> • **Quality inspectors (on the floor).**
> • **Receiving personnel.**
> • **Stocking personnel.**
> • **Shipping personnel.**

Set-up — The costs associated with set-ups or changeovers should be captured separately in order to drive the process of continuous improvement in set-up reduction. These are some of the components of cost we need to capture:

> • **Set-up operator.**
> • **Tools and equipment.**
> • **Tool inventory.**
> • **Tool room clerks.**
> • **Material lost in set-up.**

Downtime — These are the costs associated with machines which are not producing planned production parts when scheduled to do so. This element does not include planned downtime for maintenance or because of lack of demand. Here are the components we seek to capture in this area:

- **Emergency repair.**
- **Product jams.**
- **Running out of material.**
- **Breaks/Lunch.**
- **Absenteeism.**
- **Product loss.**

Maintenance (preventive maintenance) — In most World Class companies, maintenance is a category all by itself which indicates its importance. The intent of preventive maintenance is to spend more money on predictive maintenance in an attempt to save on unplanned and emergency repairs. Preventive maintenance includes all costs, such as the following, which are associated with scheduled maintenance:

- **Maintenance salaries.**
- **Maintenance supplies.**
- **Replacement parts.**
- **Product loss.**

The number of categories that a company would use in calculating the cost of production would, of course, depend on the activities and functions found in the organization. Instead of trying to give you an example of every category, we have included the most common. Any that you may wish to add would follow the same logic displayed in the categories above.

Many companies, for example, with their own sales force would attempt capturing costs associated with getting the customer to buy — from booking the order to manufacturing planning. Other companies with an extensive engineering function would look toward costs associated with bringing products to market. The intent, as we have stated before, is to understand the costs at the level at which they are created or incurred. This is why ABC selects categories which reflect the day-to-day functionality of an organization. Many of the categories, as you may have noticed, are not traditional ones, but when we take a closer look, we can see that they present large opportunities for us to save money within the organization.

Chapter Four

IDENTIFYING COST ACTIVITIES

We have conducted seminars on Activity Based Costing (ABC) around the world and everywhere we have been the participants found our presentation to be informative. Although we are quite pleased with the reception of the subject, we would much prefer that these companies implement ABC in order to gain from the many benefits which the costing system offers. The purpose of this chapter is to get the reader started on the implementation process. There is a time to learn and then there is a time to act. This is what we try to instill in all of our clients.

The ABC change process begins with the organization identifying the activities which are being performed by its resources. Resources are defined as any cost creating entity which is utilized to operate an organization's business. People, machines, equipment and even money are just some of the vital resources which a company needs to operate. Knowing how these resources are utilized is at the core of an ABC system. But knowing how they are used is not always readily apparent and that is why the identification of cost activities begins with the discussion of resource utilization.

Our intent in this chapter is to present and explain an organizational structure which uses resources as a category through which we can capture activities. These activities can then be explored to determine how their costs affect the profitability of the organization. In turn, costs are analyzed and separated into various areas of opportunity which allow the organization to exert control over product cost. The major task thus becomes finding the best way to connect activities to their appropriate cost transactions. And finally, we will show you how to use the capabilities of your present system to create the new cost system we have been describing.

Defining a Structure that Captures Cost Activities

All organizations require a structure to operate. Every company has an operating structure even if management is reluctant to publish it. The most common structure begins with an executive level at the top and then winds its way down to the lowest level of the organizational chart. In this type of structure, every task is identified with a particular place in the organization and with a particular responsibility.

Any cost system used by an organization also requires a structure within which it can fulfill its functions. Since cost is also realized from top to bottom, a cost system should be constructed accordingly in order to capture cost. If you look at the chart in Figure 4-1, you will see a cost structure which enables ABC to function at its best. The top of the structure begins with Organizational Results which breaks down further into Organizational Costs and Organizational Revenues. Revenues are captured from the products or services which we sell to our customers, but let's turn our attention to costs since they are the topic of this book. As you can see, ABC emphasizes the capturing of total costs and this is true whether the organization is principally involved in selling products or services.

To keep our cost structure as simple as possible, we have broken costs down into four levels:

•   **Product.**
•   **Function.**
•   **Activity.**
•   **Element.**

In defining product, you can think of it as an individual product, a product family or a product line. The most important concept to keep in mind is that all the costs associated with your definition of product be captured under that item.

Functions represent distinct areas which manage work within an organizational structure and therefore assume all responsibility for costs. These areas are defined differently from one organization to another. Companies call them departments, cost centers, responsibility centers or profit centers. Whatever the label, they all take responsibility for the activities which occur within their

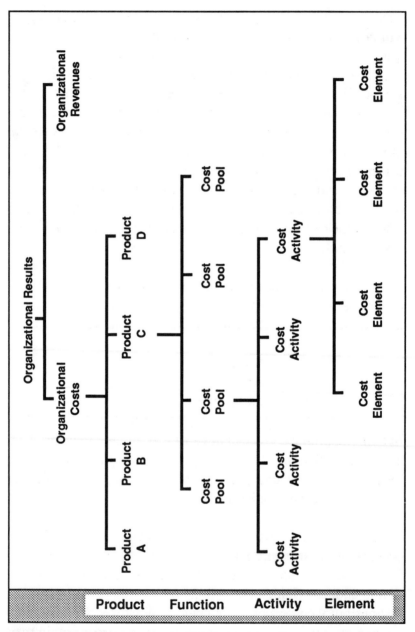

Figure 4-1

domain. The expenditures from these activities form cost pools which are then passed upward through the defined product to the organizational cost results. That passing upward is what ABC does with costs. They are driven up to and included in the product cost.

As just mentioned, companies perform various activities within each function. Activities can include such diverse items as assembling product, filling out paperwork, entering transactions into the computer or sweeping the floor. Again, the purpose of ABC is to identify these activities and to record the costs generated by them in order to drive them up toward the product cost.

Cost elements are the types of costs generated by activities being performed. These elements can be categories such as labor, material, machine maintenance, administrative work or handling costs. Each of these elements should have established parameters which define how to collect the costs associated with them.

In Figure 4-2, we have illustrated a cost structure for a familiar product to show how the costs roll up. The figure shows you how each cost level relates to each other. In addition, we can see the cost drivers for this particular product.

A cost structure such as the above not only determines the cost of performing an activity, but also shows responsibility for an activity down to its lowest level.

Transaction Processing at the Heart of All Activities

With a cost structure such as the one we have demonstrated in place, we can then begin to determine how to assign costs to actual

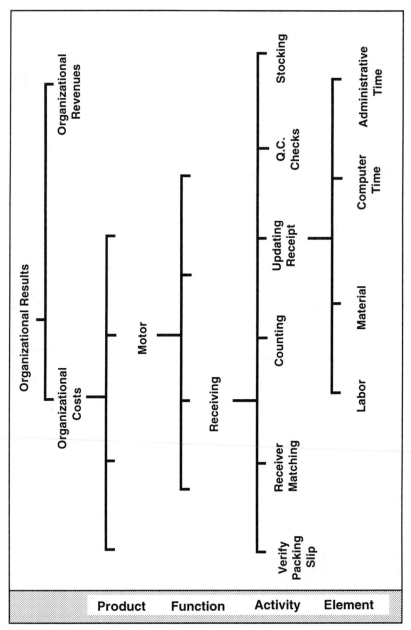

Figure 4-2

activities. Some elements like labor are obvious and what we want to do is collect data about labor in as easy a manner as possible. We don't want to make people fill out time sheets for everything they do. Pretty soon, you're filling out time sheets which record how much time you need to fill out time sheets! What is needed is a way to relate elements and cost activities by using the method your present accounting system already uses. We think that the best method to identify the relationship between activities and cost elements is to look at the transactions which are currently being processed by your system.

Every system collects transactions which it uses to update files within the system. For example, a system updates inventory by collecting transactions such as receipts, issues and completions. Or, sales files are updated by transactional information gathered from bookings, shipments and invoices. For manufacturing files, the system would use the results of operations and for financial files, it would look at payments, receivables and fixed assets. All of these transactions which are currently being captured by your accounting system can be linked to activities and these activities can be assigned a cost. Let's look at an example of how to do this:

Upon investigation, you find that for every P.O. entered into your system, a number of activities occur before and after the actual transaction. These activities are listed below:

- **Requisition Processing**
  — **Preparation**
  — **Analysis**
  — **Authorization**

- **Source Selection**
  — **New source**
  — **Contract source**
  — **Preferred source**

- **Data Entry**
  — **Multiple line items**
  — **Single line items**

- **Purchase Order Printing**
  — **Forms alignment**
  — **Forms separation**

- **Purchase Order Mailing**
  — **Stuffing envelopes**
  — **Stamping**
  — **Distribution**

Further investigation reveals that the identified activities consumed Labor and Material costs.

After benchmarking this P.O. process several times, you calculate an average cost. The total cost on average for the P.O. entry process in your company is $75.00, or the cost of the activities which go into the creation of a P.O.

All P.O.s entered into your system are tracked by transaction processing. Therefore all you need to do in order to find the total cost is multiply the number of P.O.s entered by the cost per P.O. average:

P.O.s Entered	Cost		Total Cost
125	$75.00	=	$9,375.00

In order to direct the costs back to the product level, you would separate the transactions by parts. Then you would take the transactions for a specific part used in the making of the product and cost accordingly.

The preceding example shows a company which is serious about not only identifying costs, but committed to controlling them as well. All of this investigation takes time in order to aggressively challenge total costs. Furthermore, this method of linking activities to transactions in order to determine costs will need to be applied to many other functions within the company such as the activity pools shown here:

ACTIVITY POOLS

Accounting

Engineering

Shipping

Stocking

Sales

Administration

Manufacturing

An ABC costing system relies upon the patient and persistent collection of data in order to control costs. We usually recommend in our implementation plan to investigate one or two cost pools per month. Before you know it, your company will have totally implemented ABC.

Separating Cost Reporting Activities into Areas of Opportunity

After defining a cost structure and implementing a method of collecting activity costs, you now need to turn your attention to reporting. Since the ultimate mission of your organization is to be profitable and competitive, your cost system should be aiding you by reporting areas of opportunity where you can best achieve these goals. Often there will be a number of areas and the task of finding the greatest opportunity should be assigned to a team. Some teams will determine that quality holds the greatest opportunity while others will turn to on-time delivery. Our experience shows that the area will differ from company to company.

Once you have determined the area in which you will concentrate your efforts, you will need to look at how best to display and distribute the results of your transactional processing investigations. The following key factors should be considered in the reports' format:

- **Separation of value-added costs from nonvalue-added costs.**

- **Specification of the level of reporting.**

- **Identification of responsibility.**

When reporting results, you should be careful not to display negative results by themselves. You should relate them to more favorable results so that those people who are assuming responsibility for improvement will be encouraged. The best way we have come across is to display this relationship as value added versus nonvalue-added.

The level which receives the report is an important consideration in ABC. The reported results must make sense to this level and they must be distributed to that level for a specific purpose. Don't bury your people under mountains of reports that don't in some way work toward continuous improvement. Reports should coincide with the needs of management, the requirements of supervisors and the performances of the work force. Departmental scrap, for example, doesn't mean much to a machine operator who is responsible for only one of the many operations in the department. A report on departmental scrap is made up of results from all of these operations. An operator only needs to see the results of his performance in his particular area of responsibility.

The final task of the team is to determine who is responsible for the reported results. Our experience has been that, unless you assign responsibility for each reported performance, the results will most likely remain the same or get worse. When we speak of responsibility, we don't mean that you appoint a scapegoat. Instead, you should make a person responsible for the area being measured. Then, with the reports in hand, you can analyze and discuss performance. The goal then is to determine a plan for improving upon the results. By the way, this system of assigning responsibility is valid throughout the organization and not just on the shop floor.

Interfacing Existing Systems to Create New Cost Structures

Many of our clients begin to implement ABC with the belief that they need to create or purchase a new software package. We believe, however, that you will get better results by modifying your current system since much of ABC is simply an expansion of its capabilities. Often, the principal adjustment you will need to make is the creation of new reports or the addition of new accounts. It is definitely possible to interface ABC requirements with your current system.

On the other hand, we do not recommend that you change your current system of financial reporting. Much of these results, as the list shows, are reported for external purposes.

• **Auditors**
• **Income tax**
• **Stockholders**

Your current system should be satisfying these external requirements and should not be changed in any manner. ABC should be strictly an internal reporting system. Auditors do not need to be aware of internal reports which you use to more accurately reflect what is the real total operational cost. Of course, there is a relationship between internal and external reporting, but internal reports are what you need to help run your business.

There are two other tools which are helpful in reporting ABC cost information. They are a good report writer and an allocation process. With a report writer as part of your software system, you can produce useful ABC reports which do the following:

• **Extract from financial files.**
• **Define more reporting levels.**
• **Allow for flexibility in account structures.**

An allocation process will allow management to calculate cost elements not currently appearing on performance reports. Allocations should provide for the following:

- **Splits and spreads.**
- **Statistical records.**
- **Consolidations.**
- **Financial and nonfinancial accounts.**

The only other feature which we think is nice to have for an ABC system is the capability of producing graphics. Graphics are more easily understood since there isn't the need to absorb the meaning of rows and columns of numbers. And better understanding of how our companies run is at the root of ABC. In the next chapter, we will explore even further how to develop this new costing methodology.

Chapter Five

DEVELOPING THE PRODUCT/SERVICE COST EQUATION

Let's return to the baseball metaphor with which we opened this book. Imagine a third-base coach trying to decide whether to wave a base runner home with the winning run in the last inning of the final game of the World Series. A bad decision on his part could cost his team the game and the championship. Something of the same can be said about traditional cost methods. They are not only inaccurate, but they can be detrimental to your business. They create or collect bad data and bad data leads to bad decisions.

Since every decision we make has an effect on the bottom line, we must start with accurate data in order to feel comfortable in making the decisions which direct the operations of our organizations.

In organizations which are seeking to become World Class companies, it is imperative that problems and issues are evaluated and solutions are implemented through some problem-solving method. More and more today, such a method entails the use of some form of team play. In his book, **People Empowerment:** *Achieving Success from Involvement,* Pro-Tech vice president Wayne Douchkoff shows how to develop such teams and how to make them work. "The mission of every company," he says, "is to *involve and empower people* to solve problems and find areas of opportunity."

Solving the major problems associated with inaccurate data lends itself to this multidisciplinary team approach. We suggest the formation of a financially-led team which is challenged with the task of improving the accuracy of the cost data used to support the management decision-making process. In this chapter, we are going to focus on the development of the total product or service cost equation. We will then discuss the components of the equation and suggest the proper mind-set for its most effective use. But, it all begins with accurate data. We cannot stress this enough. The better the data, the better the information. Peter Grieco has consistently maintained that many companies are "information rich and data poor."

Converting Indirect to Direct Cost Methodology

Indirect cost has long been the mystery factor in most cost equations. Little effort has been expended by management to associate these indirect activities with the actual products or services that utilize them. This is primarily due to the nature of indirect activities. They usually involve functions of an organization which are used to support production rather than the actual production of products. Thus, they are perceived as being too difficult to separate so that they can be applied to a single product's total cost. Some examples of the more familiar of these functions and their activities are listed below:

FUNCTION		
MANUFACTURING	SERVICE	ACTIVITIES
Material Handlers	Mail Room	Delivery of materials/ data to various work centers and operations.
Inspectors	Data Entry	Perform various quality audits throughout various work centers and operations.
Supervisors	Supervisors	Support various products and services being produced.
Production Planners	Customer Service	Plan for the activities of various products and services.
Maintenance	Service Dept.	Perform machine and facility repairs and preventive maintenance.
Cycle Time Mfg.	Cycle Time Admin.	Perform changeovers on various lines, facility equipment and elements.

The list, as you well know, goes on and on. Perhaps that is part of the reason why the mystery of how to collect or charge these costs more directly continues as well. This is precisely where Activity Based Costing comes into play.

Activity Based Costing focuses on the activities being performed and then determines the best way to directly charge or allocate more realistic proportions of costs to various products and services. At the same time, every effort is made to determine methods which *reduce the amount of data that must be input by individuals*. Effort is put into relying on calculations which support the total cost equation. Let's look at some examples of just what we mean, using the functions and activities identified in the chart above. In the examples which follow, we will demonstrate how to use methods which more directly charge costs that are usually considered indirect.

<u>FUNCTIONS</u>	<u>ACTIVITIES</u>
Material Handlers and Mail Room	**Allocate based on relationship that best represents an equitable distribution of effort among products and services. We recommend determining the factor based on a relationship between the number of components moved or mail delivered and the number of products produced or services provided. This factor can then be applied against the total expense to include labor and equipment.**

Example: Total Cost = $7,879 per month

Product	Product Built	Number of Components	Total Moves	Ratio	Product Cost
A	300	750	225,000	8%	$ 630
B	75	1,000	75,000	2%	$ 158
C	5,000	480	2,400,000	89%	$7,012
D	25	100	2,500	1%	$ 79
			2,702,500	100%	$7,879

Figure 5-1

FUNCTIONS
Inspectors/Data Entry

ACTIVITIES
Should always be collected directly as a component of cost. Quality checks at operations should be analyzed as you would a labor operation. It should be estimated and become a part of your standard.

FUNCTIONS
Supervisors

ACTIVITIES
Allocated based on productivity of products produced or services provided under their charge. Productivity can be determined by the standard hours expended by products.

Example: Supervisor's Costs

Supervisor	Salary (Daily)	Products	Standard Hours	Ratio	Costs
#1	$96	A	400	53%	$50.88
		C	300	40%	$38.40
		D	50	7%	$ 6.72
			750	100%	$96.00
#2	$120	A	350	32%	$38.40
		B	750	68%	$81.60
			1100	100%	$120.00

Figure 5-2

FUNCTIONS	ACTIVITIES
Production Planners and Customer Service	Allocate expenses based on orders entered for products to be produced or on purchase activity.
Maintenance and Service Dept.	Direct charge of dedicated repair or maintenance services. Services not dedicated should be allocated based on any applicable basis — volume, hours, orders, etc.

FUNCTIONS	ACTIVITIES
Cycle Time Mfg. and Admin.	Direct charge to the product or service for which the changeover is being performed.

No method of allocating will satisfy all indirect charges. Instead of trying to find the perfect method, we encourage you to be creative in selecting an equation that will provide the best method for improving the accuracy of data. Once such a method or equation is established in your company, every effort should be made to have your present system perform the necessary calculations on whatever schedule you choose, either monthly or weekly. The goal of cost management should be to eliminate indirect charges and to allow the computer to make as many direct charges as possible.

Examining the Relevance of the Components of Overhead

In our consulting work or at our seminars, we often quiz cost managers on the relevance of overhead. Some just smile and say nothing. Others hem and haw and try to justify their present practices. Some even refuse to discuss the issue and walk away. A few will admit that overhead may be irrelevant, but they see no other alternative but to use it. Some people even tell us they are hard at work trying to eliminate overhead altogether.

We believe that overhead does serve a purpose in the standard cost equation. Our concern, however, is that a great number of the components of overhead do not belong there. In fact, many of the components are waste disguising themselves as overhead so that they won't attract our attention. Companies who are involved in World Class programs of improvement are focusing on waste reduction. Cost management in such companies is also involved in identifying the waste components which alert companies to opportunities for reduction.

Overhead must be analyzed to determine its components so that you can identify what is wasteful. And what is waste? This is the definition that we use:

WASTE

**Anything other than the minimum resources
of people, machines, or materials
that add value to products.**

The components of overhead differ from company to company, but the typical definition of overhead is all costs other than direct materials and labor. It is no surprise then that overhead is the biggest source of cost in most manufacturing operations. Are these costs relevant? The answer is yes. They affect product cost which affects margins which affects profits which determine the success of the company. What we are saying is that the components of overhead which can be identified as adding value should be charged directly to the activity. If the components do not add value to the activity, they should be eliminated. All costs in a company should relate to activities which produce or support the products and/or which the company sells.

Take a moment right now and consider the activities performed in your organization that ultimately wind up being charged to an overhead account. Start by thinking about a specific function such as sales, engineering, quality, finance, maintenance, materials management, or shipping. Are the activities within a specific function relevant to product cost? If they are, then they need to be evaluated and intelligently distributed back to the products which utilize these activities. As we advised before, we recommend forming a team of people who will take their clipboards and visit the targeted function. There, the team will follow the quality road map of data gathering and investigate the activities being performed to see which ones are relevant to product cost and which are not.

We are always surprised by companies that fail to do this. Almost all of them use teams to investigate activities and operations on the manufacturing floor, but, for some reason, they never go out and challenge administrative and indirect functions. Activity Based Costing looks to challenge any cost producing activity within the organization.

Let's use the Purchasing function of a company to demonstrate what we mean. Here is an example of how one Purchasing function may conduct its day-to-day activities:

- **Phone Calls — Internal and External.**
- **Requisition Processing.**
- **Retrieving MRP Order Action Reports.**
- **Entering Purchase Orders.**
- **Distributing and Filing Paper.**
- **Mailing Orders.**
- **Expediting Activities.**
- **Data Base Maintenance.**
- **Meetings — Internal Planning.**
- **Meetings with Suppliers.**
- **Performing Surveys and Audits.**

These activities will, of course, be different from company to company, but the above are typical. The best way to determine what activities are present in your company is to observe the Purchasing function in action. After the list of activities is completed, the team then decides what resources were involved in completing the activities. The resources which are utilized in performing a functional activity are the cost elements which we need to identify, capture and distribute to the products. A list of such resources would consist of the following items:

> - **Labor — People performing tasks over time.**
>
> - **Communications — Phones, fax machines, EDI lines, telex machines.**
>
> - **Computer Utilization — Data entry, maintenance inquiry, printing.**
>
> - **Equipment — Typewriters, calculators, mailing machines, copiers.**
>
> - **Supplies — Office supplies, forms, folders, boxes, pens, pencils, pads.**

Each of the above resources generates costs which can be associated with the Purchasing function. The following expenses must also be recognized and captured as well within this function:

> - **Travel Expenses.**
> - **Facility Expenses.**
> - **Utilities Expenses.**
> - **Acquisition Expenses.**
> - **Equipment Expenses.**
> - **Communications Expenses.**

As you can see, the ABC system looks to identify the total cost of running the department. Many of the cost elements that we have identified can be captured directly, while others will need to be allocated according to some ratio that you determine. The system we have demonstrated here is much different from traditional systems in which all purchasing costs were realized as overhead. Once they were hidden in overhead, they no longer were relevant to product or administrative costs. ABC seeks to change this attitude to one in which all costs are either made relevant to products or eliminated.

Utilizing the Capabilities of Cost Allocations in Cost Development

As a part of their present cost systems, most companies have the capability of allocating various expense elements collected in the aggregate to detail levels which allow for a more equitable spread. This spread, however, is normally performed on nonoperational costs and not on product costs. Expenses such as rent, utilities, maintenance, property and facility taxes are among those common allocations which are spread to the department or cost center level.

Activity Based Costing seeks to go even further and extend the allocation process of expenses like those just listed to the product cost level. An allocation process which effectively *splits and spreads* manufacturing and nonmanufacturing expenses back to the product level should be the driver within an ABC system.

A *split* in the ABC system is defined as *the process of identifying and combining similar expenses from the general ledger for the purpose of distributing cost to more operational categories (e.g., cost of quality, cost of inventory, cost of selling, etc.).*

A *spread* is defined as *the process of reallocating selected categories of expenses to product levels for operational reporting, analysis, and problem solving.*

It is important that Cost Management use such capabilities to develop operational reports which will help the organization focus on nonvalue-added cost and wasteful activities which are eroding margins. An ABC allocation system will greatly enhance the ability of the Cost Manager to produce product cost reports which are more realistic, appropriate and flexible in order to meet the expanding needs of operational reporting. There is an issue of practicality here, however. Collecting data can be expensive. As we try to get all cost data more directly reported to the product level, the cost of collecting data may equal the amount of savings we expect to realize.

An ABC system recognizes that not all expenses can or should be allocated directly to the product level. However, every effort should be made to allocate as many expenses as is logical, justifiable and practical. Our discussion on allocations which follows will stress the ability to collect and distribute cost data into reports which support management's objectives. The allocations will reflect valid cost reducing opportunities.

The allocation process begins with an understanding of where the data comes from which we are planning to utilize in order to enhance the credibility of reported results. The following list describes a number of these sources:

General Ledger — The primary source of data for allocations is stored in the general ledger. Account totals for a specific expense or a combination of accounts can be split off the ledger and used in the process as details.

Statistical Record — Statistics can be generated separately by the Cost Manager in order to enable the allocation process to perform calculations based on information not available in the General Ledger. Statistics can include numbers such as head counts, facility square footage and transaction counts or ratios such as percent of effort or waste.

Previous Allocations — Allocations can be stored and used in performing additional allocations throughout the process. An example might be the allocation of machine utilization to products levels which could then be used to allocate machine maintenance to the same product.

Budgets/Forecasts — Budget and forecast files can be used to compare their figures with actuals and to build output results or ratios which can be used in additional allocations.

Transaction Files — Transaction files allow the allocation process to develop ratios by capturing the number of times an activity or function occurs and using the ratio to spread the cost back to products more equitably.

All or part of these sources can be used to allocate cost to product or service levels. It is the job of the Cost Manager to look for any source of data that can be used to help in the calculation of more accurate and visible results. Furthermore, the allocation process should allow the Cost Manager to access these data sources internally within the present system and not externally through some manual input such as journal vouchers to the general ledger.

The cost system should meet the needs of the user and not the other way around. What we are talking about can be seen in the flow chart below of a flexible ABC allocation system.

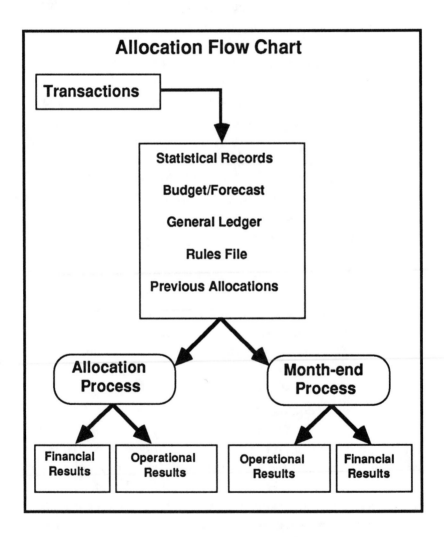

Fig. 5-3

Notice how the ABC system is integrated into the month-end process. Unlike some other systems, it is not an afterthought which is to be accommodated some time after the month-end results are generated. ABC strives to make results available on a timely basis so that both financial and operational results are available to make business decisions.

The system above is driven by certain rules or parameters which identify the logic needed to perform the necessary calculations for each allocation. These rules should be kept on file and should be easily accessible to the process, if not part of the computer system itself. Each company will have various kinds of rules in its system depending on its needs and degree of flexibility. The list of rules which follows will give you some idea of what is necessary.

> **Allocation Sequence** — Allows the system to determine which set of allocation logic to run and when. By using a sequence number, the system can perform allocations based on previous allocations which were performed in prior runs.

> **Organization From/To** — Identifies to the allocation process that an inter-company allocation will be made across legal entities. This parameter must be followed by an offset account to ensure that the system will keep the "from" and "to" companies in balance. When allocation is taking place intra-company, this parameter is not necessary.

> **Offset Account** — When the allocation process crosses legal entities with data, an offset account must be used to keep the books in balance. If allocations are only being made in the same legal entity, then an offset account is not needed.

Start/Stop Accounts — Identifies the account
number from which to start collecting costs and
the account number at which to stop. This allows
the system to add a string of contiguous account
balances together without listing each account
separately. If the accounts are not numbered
consecutively, then the user should be able to list
each account separately.

Allocation Method — Identifies what method will
be used in calculating results. The system user
should be allowed to be as creative as possible as
long as the results are realistic.

Journal Type From/To — When allocating, the
system should be able to selectively source and
update from several different files present in the
system. Data can be captured, calculated and
then updated in the selected journal files, such as
the general ledger, budget, forecast or statistics
files.

Percentages/Amounts — Depending on the type of
allocation method, the system should allow the
user to predetermine fixed ratios or amounts to be
used in calculating allocations.

With a set of data files in place and rules to control the process, the
user is now ready to utilize both aggregate totals to detail cost
factors in product cost development. Let's look at an example of
how the allocation system we have described will work.

Example: **Allocating the Cost of Purchasing**

The expenses associated with the purchasing function have been identified as salaries of buyers and clerks, supplies used by the department and all other expenses. The miscellaneous category is traditionally captured as part of overhead.

Activity Based Costing attempts to get closer to the true total cost associated with the purchasing function by allocating overhead costs back to the product level. These costs then become part of the cost factors associated with the process of buying materials utilized by the entire organization.

What cost factors are associated with the purchasing process?

To answer this question, you need to do some research to find out what are the activities that occur on a daily basis within Purchasing. Some of these activities are obvious, such as salaries of personnel in the department and those supplies which have been charged back as expenses. However, some expenses (as we have seen) are not so obvious and do not get collected by the cost system. Let's look at how some of these costs can be obtained through an ABC allocation process.

> **Utilities** — base costs on square footage.
> **Computer costs** — base costs on transaction volumes.
> **Communications costs** (telephone, fax machines, etc.) — base costs on connect times.

> **Equipment costs** — base costs on depreciation schedules.
> **Copier costs** — base costs on number of documents reproduced.
> **Acquisition costs** — base costs on number of activities taken to procure products, externally or internally.

> Each individual cost is then accumulated into an identified account, such as the cost of purchasing total, whereupon it can be spread back to specific products.

> **How can the Cost of Purchasing Total be distributed back to specific products?**

> The allocation of costs back to products should be based on an intelligent method which relates products to the identified costs. In Purchasing, this relationship can be established through the number of purchase orders or the number of transactions per product. Both of these relationships are based on the effort expended to purchase materials for products. In order to determine the percentage of total cost which should be allocated to a specific product, first determine the ratio of the number of purchase orders (or transactions) to the total number of purchase orders (or transactions). That ratio will be the percentage for you to use.

Activity Based Costing uses allocations like the ones described above during month-end processing. However, there is no reason why allocations can't be run on a daily basis as part of the nightly processing. The objective, after all, is to provide the organization with the results of operations as quickly as possible in order to enable the analysis and problem-solving effort to function with the latest possible data.

Not every computer system has the capabilities we have mentioned in this chapter. Cost managers will need to make the most out of the allocation capabilities which do exist. If that proves inadequate, they should be out shopping for a new software package. Consider what your needs are and then look for a system which supports World Class activities. Be careful, however, of systems that say they can do it all. ABC requires a lot of input about how your operation functions. No system by itself can "know" all that information.

Expanding the Elements of Product Cost to Reveal Product Line Profitability

The key to product profitability is to eliminate the components of cost which don't add value to the product. If the product still remains unprofitable after these actions, then it should be dropped from production. Unfortunately, this sounds easier than it actually is. Many companies do not have cost systems which accurately reveal the profitability of a product. This is often because costs are collected and reported in the aggregate, so that one product's costs are contained within a lump sum for all products, some of which are profitable and some of which are not.

Aggregate costing tends to unfairly represent the true elements of product cost. Some products suffer unfair assessments of overhead cost and some not enough. Activity Based Costing attempts to eliminate this unfair distribution of cost to allow only applicable elements in their proper ratios to be charged to products. At the same time, ABC evaluates activities which don't add value to products and seeks to reduce or eliminate them.

Those cost activities which are generating waste need to be made visible in order for the improvement process to identify root causes and to consider alternative solutions. ABC helps to expand the list of cost elements in product cost reporting in order to gain this increased visibility. Let's look at an example of an expanded list of cost elements:

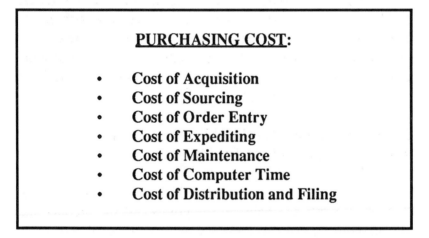

All of these elements of cost need to be reviewed and challenged on a regular basis. At the very least, there exists an opportunity to reduce the effort and resources used in each of the elements above.

Although we looked at only the Purchasing function in this chapter, a company should undergo a similar research and analysis process for all the functions within the organization in order to assist in the cost reduction effort. The key to cost reduction is making visible what the true total costs are in running an organization. When companies are exposed to what these costs are and how they affect the products they build, we believe everybody in the company will make a concerted effort to contribute to the reduction.

Chapter Six

COST MANAGEMENT'S RESPONSIBILITY IN THE WORLD CLASS AGILE MANUFACTURING ENVIRONMENT

The role of the cost manager will expand in the World Class, agile manufacturing environment. We will not only look to them to produce results and reports but to get involved with the improvement of results. The financial manager will need to play a key role in determining how management objectives are measured. He or she will need to lead an effort to identify the elements of cost and to trace them back to reporting objectives. Cost management must also ensure that results are being distributed to the proper levels of the organization for decision making and corrective action. The impetus behind all of these efforts is to encourage participation in order to reduce or eliminate costs throughout the organization.

In other words, cost management's responsibility can no longer just be defined as the collecting and reporting of operational results. In today's business world, the success of an organization will undoubtedly depend on a more aggressive level of involvement by cost management and on their aggressive containment of costs. The time has come not only to process the numbers to come up with results, but to find out what is behind the results. The time has come to question the importance of the results we produce, to ask ourselves questions like the following:

- **Who is getting operational results?**
- **Which functional area is achieving operational results?**
- **Does the organization understand the results?**
- **Are the results and measurements being used to improve the operation?**
- **How relevant are the results?**
- **Do the results reflect daily operational reporting?**

Notice that the above questions demand participation and involvement by all levels of an organization. We can no longer tolerate an attitude in which no one cares or takes responsibility for the results. In the agile organization, we can't afford to hear somebody say: "All I do is prepare the results. If nobody uses them, it's not my fault." With an Activity Based Costing system, it is imperative that you develop a structure for using the results. Do people understand them? Are they relevant?

This chapter deals with the involvement and responsibility of cost management in an environment faced with the demand for continuous improvement. That is the situation facing most businesses today. We want to provide each organization with a methodology showing how to focus their improvement efforts on areas and activities that will offer the greatest benefits. Such a methodology has been a long time coming. Our process can demonstrate where to focus by implementing a decision-making process which is grounded in data that realistically reflects opportunities, goals and benefits. And just recognizing these opportunities is not enough. Management must see to it that involvement and responsibility are assigned throughout the organization. They must also make sure that whenever a project begins that the team addresses the issue of start and stop dates and assigns definite times. It is the responsibility of cost management to provide the organization with the data that benchmarks opportunity, establishes realistic goals and displays the costs versus the benefits.

Cost Implications of Agile Manufacturing

Focusing on improvement also means looking toward the future. A proactive stance is a necessity in today's business world. Companies must begin considering the cost implications of manufacturing in the 21st century. On the horizon now is a vision of manufacturing in the next century known as agile manufacturing. Many of you have already heard of Japan's quest for the three-day car. The production of a car in three days is an example of their version of agile manufacturing as developed by that country's Japanese Manufacturing 21 Project. Some other examples closer to home are General Motor's Saturn, Benetton and Wal-Mart.

AGILE IDEAS

SATURN

In planning production, Saturn "images" or creates stock orders for its dealers. The retailer then has the opportunity to change the order in real-time according to the desires of its customers. The new specifications are directly reported to production scheduling at the Saturn plant.

BENETTON

Instead of dying yarn and then knitting the sweaters, Benetton produces finished sweaters in neutral colors and then dyes them to meet the market demand for colors.

WAL-MART

Wal-Mart lets individual stores order directly from suppliers. Using this method, Wal-Mart maintains high service standards with 25 percent of the inventory. The company has also been able to cut restocking time from 6 weeks to 36 hours.

What we are experiencing is a paradigm shift from the lean and flexible styles exemplified by Just-In-Time and Total Quality Control to a more inclusive style which emphasizes the operations of the entire company and not just the factory floor. We have summarized these shifts in Figure 6-1.

AGILE MANUFACTURING

The Paradigm Shift

Lean/Flexible	Agile
•Eliminate inventory	•Zero inventory
•Eliminate waste	•Zero waste
•Flexibility in scheduling	•Build to sales — daily
•Shortened lead times	•Minimum lead times
•Six Sigma quality in products and services	•Quality & reliability measured in terms of total life cycle costs
•Low unit cost via large volumes of similar products	•Low unit costs from:
	-Modular production facilities
	-Easily programmable equipment
	-Enterprise integrated information systems
	•Virtual products for each customer
•Focus is on factory floor enhancement of thru-put	•Focus is on total enterprise cycle time
•Task-oriented training of employees	•Enterprise-based training of employees for maximum capability and creativity
•Equipment & technology as primary asset	•Employees as primary asset
•Effective use of resources to contain costs	•Social responsibility
	-Products designed for recyclability and reconfigurability
	-Design-focused product changeover capability

Figure 6-1. (Part 1) Agile Manufacturing—*The Paradigm Shift*

AGILE MANUFACTURING
The Paradigm Shift

Lean/Flexible	Agile

•Clearly defined roles
  -Customers
  -Competitors
  -Suppliers
  -Manufacturers/providers
  -Stakeholders
	•Constantly changing roles as
	  defined by the requirements
	  of the virtual enterprise

•Broad-based market view:
  economies of scale
	•Economies of scope: focus on
	  servicing ever smaller niches

•Fragile to the impact of change:
  optimized for one purpose
  -Focused factory
	•Change optimized:
	  -Equipment
	  -People
	  -Information systems
	  -Equipment
	  -Supply base
	  -Admin. systems & structures
	  -Technologies

•Product designs are rigid and
  "frozen" only after numerous
  changes and enhancements
  -Value-added approach
	•Products are designed for:
	  -Producibility
	  -Maintainability
	  -Disassembly
	  -Reconfigurability
	  -Upgradability
	  -Recyclability

•Products are designed for
  internal integration
	•Products designed for
	  maximum cycle time
	  effectiveness

•Operationally focused:
  -Short-term financial
  -Extend status quo as long as
    possible to amortize costs
	•Strategically focused:
	  -Long-term performance
	  -Diffused authority
	  -Dynamic corporate structure

Figure 6-1. (Part 2) Agile Manufacturing—*The Paradigm Shift*

Agile companies will need an accounting and finance system which assigns costs more accurately to products and services and which provides management with the cost information needed to make strategic decisions. Activity Based Costing will be the system to meet these needs. In the future, there will be an even greater need to establish time-based metrics to monitor your progress. For example, design engineering would measure themselves on their progress in these areas:

> - **Total design cycle time.**
> - **Number of engineering changes.**
> - **Material costs.**
> - **Labor costs (direct and indirect).**
> - **Production cycle time.**
> - **Test cycle time.**
> - **Components per product.**
> - **Component standardization.**
> - **First pass drawing accuracy.**
> - **Number of preproduction and prototype units required.**

Movement will also need to be made toward lowering costs by integrating information systems and designing equipment so that set-up time simply involves reprogramming the machine. There should be no physical removal of tools or dies, for example. The machine would reconfigure itself to the new job according to the instructions it receives from the company's integrated data base. This cost-cutting would also be reflected in the design of modular production facilities and products. Some futurists see the possibility of buying one car in your lifetime. As parts wear out or get improved, they would be replaced at "service centers." Even if you wanted a style change, your car would be outfitted with new body panels.

All these developments on the horizon will lead to a radical change in manufacturing which will require companies to adopt a service business strategy. This new business style will place an emphasis on zero inventory, zero waste and, perhaps most importantly, a fast response time and short production cycle. None of this is possible without accurate, timely and *relevant* cost information. Activity based costing will be the tool in use as we go through the paradigm shift to agile manufacturing.

Expanding the Role of Cost Management's Involvement in Operations

The best way to determine the importance of being involved in operations is to walk through the operational areas of the company. This is often called MBWA, or Management By Walking Around. Assess the people and processes utilized, areas in which they work and the degree of supervision they receive. Ask each person how they are measured. Do they know? Do they receive performance reports, graphs or targets to work toward? Unfortunately, the answers to these questions are all too often one "no" after another. How can we expect improvement in this type of environment? There is a serious need to provide operators with performance measurements that will propel the improvement effort. Cost management should play an instrumental part in providing the operators with the data they need to do their job better.

Correcting this deficiency in data distribution is one of the first efforts an organization needs to make in becoming a World Class company. We have developed a five-step improvement process which we use in our consulting work. We call these five steps a company's "nickel defense" against waste.

THE NICKEL DEFENSE
A FIVE-STEP IMPROVEMENT PROCESS

The **first step** is to get out of the office and into the places where a company's operations are occurring. Look for areas of opportunity. Identify waste and take notes.

The **second step** is to review your findings and establish a method that will allow you to assign a cost to each wasteful element.

The **third step** is to start collecting data that will allow you to begin measurement. Be sure that the data is as accurate as you can make it so that the data is believable.

The **fourth step** is the most important. Educate people about your plans for performance measurements. Make sure that they understand what the measurement is saying and how their tasks can affect the results being measured.

The **fifth step** should be to make the process an ongoing one. Measurements should be reviewed with operators to ensure that their performance improves and that waste reduction does indeed become a reality and not just a wish.

Cost management must also be involved in training itself and the rest of the organization to identify waste that negatively affects company performance. There is no better way to learn where this happens than to get out of the office and to gain the support and participation of people. Don't, however, go out and start badgering them. Asking a "why" question should never be threatening. It is a much better idea to present the question as a means to understand an issue.

After you have asked the question, make sure that you are truly listening. Summarize data and information about each issue. We have found that the best way to get this information is to use a cross-functional team upon which cost management is represented. The team can set up a structure for the investigation of waste in which efforts are not duplicated or opposing each other. The team's mandate is to look for opportunities for improvement and to seek out the signs of cancer which are eroding profits. Here are some of the signs of opportunity that a team should look for:

SIGNS OF OPPORTUNITY

- **Customer complaints about the products delivered or services rendered.**
  - **Quality**
  - **Delivery**
  - **Quantity**

- **Product not produced or orders not booked until the last week of the month or quarter.**

- **Machines that should be utilized are down for repair.**

- Significant levels of inventory throughout the operation.

- Major out-of-stock situations exist.

- Finished goods are not being shipped because they are waiting for documents, sales orders or invoices.

- Obsolete inventory is being identified, however little or no action is being taken to eliminate it.

- Excessive set-ups/changeovers are creating large lot sizes instead of approaching the ideal lot size of one.

- Employee surveys reveal frustration and mistrust.

This list could go on and on, and, in fact, it should as a team identifies issues, and thus opportunities, facing the organization. Each of these issues is costing the organization, either loss of competitive position or profits. As we discussed in Chapter Two, you must include the cost of nonconformance as well as conformance when figuring the cost relationships of the issues you uncover.

Computing Benchmarked Costs and Distributing the Results Internally

The success of any improvement program must be measured financially. In order to accomplish this, cost management must be able to compute a cost for each of the areas which it is benchmarking. Benchmarking requires that a company look at both internal and external activities and compare those results with the results obtained by World Class companies. The cost data which you collect becomes the starting point against which the improvement process is measured. The computation of these costs must be accurate, but that does not mean that the computation must be done in a traditional manner. Performance measurements, in many instances, will not be governed purely by financial parameters. Often, as we have shown in previous chapters, there may be parameters that we never considered or calculated in the past.

Even though we should strive to make data as objective as possible, there are times when subjective data may be required to establish a measurement. Let's look at one example:

An organization is experiencing an excessive amount of material stock outs. Since this has been occurring for a number of years, the company has now become very proficient at reacting to stock outs and has developed a process to deal with them. They put aside any product in process and start production on another product until the stocked out material is available to complete the sitting product. Since the company experiences this situation so frequently, it is now thought of as being normal. Thus, nobody has aggressively worked to resolve the issue.

Does this sound familiar to you? Can you think of an issue handled in a similar way in your organization? Let's analyze what is really happening and present a way to begin the improvement process.

The first question to ask is:

How often does this issue occur?

Often the issue comes up so often that it is perceived as being an elephant-sized problem. If that is the case, then look at a smaller piece, such as just one product, part or operation. For example, take a look at one product that experiences frequent stock outs. Then the team should determine a way, automated or manual, of keeping a count of the number of times a stock out occurs. This data must be objective and represent a true picture of the number of occurrences in order to be effective. We prefer to use the Pareto chart format in this instance.

The second question is:

How many elements or activities of cost will this issue affect?

There are many elements of cost which can be affected by this issue. At this point, you should not hesitate to use subjective data when real data is not available. Remember that the results that you compute are used internally. The only criteria they must meet is that they challenge your organization to improve. Let's look at what some of the elements of cost could be for our example:

- **Cost of material which is sitting.**

- **Cost of the labor in the product which is sitting.**

- **Cost of product not being produced (opportunity cost).**

- **Cost of expediting the stock out.**

- **Cost of unplanned changeover.**

- **Cost of management or supervisor involvement.**

- **Cost of time lost.**

Your team will need to brainstorm the elements which are applicable to your organization and your particular problem, but the list above gives you an idea of how to go about this task. We suggest using a fishbone diagram at this point to explore causes and effects.

The third question is:

How do we determine a cost for each element or activity?

To determine the cost of an element or activity, be prepared to use both financial and nonfinancial figures. Incomplete products taken out of production have accumulated material and labor costs. These values should be costed along with an inventory carrying cost for each hour or day the product sits waiting to be finished.

An expediting cost should also be assessed for all of the resources used to get delivery of the stocked out material, including the items listed on the facing page:

- **People cost.**

- **Phone cost.**

- **Computer cost.**

- **Freight cost (priority or regular).**

- **Handling cost.**

- **Equipment cost.**

Changeover costs are usually a function of time and would involve any activity necessary to reinitiate the production process when parts become available again. When calculating these costs, consider both machine set-up and preparing work areas to begin production again.

When determining costs, don't forget to include the cost of having management and supervisors spending their time getting involved. Don't overlook any time that may have been lost making a decision on what action to take. This cost is related to the lost opportunity of making other products.

The fourth question is:

How do we calculate the cost per occurrence?

Based on the computations you make in the previous step, you would begin to record a list of the costs associated with a stock out.

Inventory carrying cost

$$\text{I.C.C. } \times \text{ value } \times \text{ days}$$
$$.0008 \times 3,750 \times 3 = \$9 \qquad\qquad \$\ \ 9$$

Expediting charge

People — standard rate $\times$ time spent
18.45×15.5 hrs = \$286

Equipment — standard rate $\times$ time spent
Computer $13.50 \times 45/60 = \$10$
Phone 4.75×3 hrs = \$14
Handling $22.75 \times 30/60 = \$11$ \$ 321

Freight (Priority)   \$ 44

Changeover cost

standard rate $\times$ time spent
$18.45 \times 30/60 = \$9$ \$ 9

Time lost

hours lost $\times$ value
6 hrs $\times$ \$345 = \$2,070   \$2,070

Total   \$2,453

The calculations above are based on evaluations made by the team on several different occasions in order to arrive at an average time spent and an average amount of resources utilized. Weighted averages can be used when the part in question has a substantial difference to average.

The fifth question is:

How do we associate the cost to the occurrence?

Once the cost per occurrence is determined, we then multiply that figure by the number of occurrences to arrive at the cost of stocking out.

Average cost X # of occurrences
$2,453 X 15 =   **$36,795**

The cost of stocking out should then be reflected in the cost results as a monthly cost. Whether your calculations turn out to be more or less is not important. What is important is that you use this cost as your benchmark when establishing goals and objectives.

The sixth question is:

Who should be exposed to the results and
who should be held responsible for reducing the cost?

It is up to the team to decide the best way to distribute cost results. Every attempt should be made to relate the measurement to the activity so that the individual receiving the results has relevant data from which to work toward improvement. Needless to say, it would be advisable to include the individual receiving the results on the team from the start.

In some cases, more than one function may receive cost results since the need for results is common to many areas requiring improvement. Caution should be used, however, when distributing results to more than one function to ensure that people understand exactly what is expected from them.

Responsibility for improvement must ultimately reside with the person to whom the cost results have been distributed. That is why these people need to participate in the understanding of the measurement's meaning, the establishment of goals, and their level of responsibility. Lastly, you should collect and monitor all contributions to the improvement of a process to ensure that these activities do not negatively impact other areas.

Supporting the Continuous Improvement Effort through Visibility of Cost vs. Benefit

As we have already noted, cost management must be involved in the creation and distribution of cost results to other functions and activities within an organization. This is necessary, not only to disseminate results, but to build support for the continuous improvement effort. The whole organization needs to be committed in order for the effort to have the best results. That is why it is so important for cost management to make sure that the cost results are in line with company objectives and that they help the company focus on reducing or eliminating nonconformance costs throughout the organization.

There are a great many types of results that can and should be created, but every improvement effort should be stated in dollar amounts. This allows the organization to calculate the costs versus the benefits for either a part or the entirety of the program. Every team and function in the company should be stating their improvement results on a cost vs. benefit basis for ease of comparability.

All measurements at the beginning of the particular effort should be set at zero. (See Figure 6-2.) This represents the fact that no dollars have been spent or saved at the beginning of the program. The other rule in developing cost vs. benefit measurements is that

all costs associated with the effort to improve must be captured
and applied to the project from the very first meeting of the team
conducting the work. That means all dollars for consulting,
training, education, meetings, materials as well as any other
expenses incurred by the team for that activity.

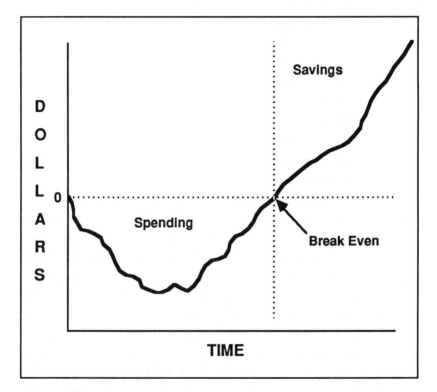

Figure 6-2. Cost vs. Benefit

The offset of the incurred expenses should be any savings resulting from the improvement effort. Depending on the project,
different types of savings will be measured to reflect how well the
team has benefited the organization. Team members should

discuss with cost accounting the cost targets which will be needed to demonstrate the level of savings. Some cost targets to consider are the following:

COST REDUCTION TARGETS
- **Labor.**
- **Material.**
- **Processing time.**
- **Lead time.**
- **Paperwork.**
- **Set-up time.**
- **Parts.**
- **Suppliers.**
- **Queue time.**
- **Move time.**
- **Wait time.**
- **Cycle time — Admin. and manufacturing.**

Even though all of a team's measurements must ultimately be reflected in dollars for a cost vs. benefits chart, this does not mean that teams can't keep measurements in other units for their own internal use. For example, they may want to keep measurements to reflect time, percentage, weight, quantity or any other unit of improvement. But, so that the whole organization can compare apples with apples, teams must translate all these measurements within the organization into dollar figures for general use. Besides ease of comparability, no unit of measure can be as sensitive as dollar amounts. Management understands dollars. Employees understand dollars. And everybody understands that aggressive work to improve an activity will eventually pay off in dollars.

As we previously mentioned, any improvement made within the organization will ultimately also have an effect on the bottom line.

Management must come to understand that nontraditional and nonfinancial measurements need to be considered as well. Activity based measurements are used to provide management with the detailed results necessary to identify areas in need of improvement. All the measured results in the world amount to nothing, however, if programs to improve are not initiated. An unattended or unmeasured problem does not disappear.

Providing the Data for Investment Analysis and Management Decision Making

Developing an investment plan to support an organization's future challenges a business executive to go beyond traditional practices. A cost management system must be called upon to provide more detailed data than normal in order to help with the decision-making process. Furthermore, this data must not only reflect the finances needed to compete in the global marketplace, but reflect the need to invest in the improvement process which we have been discussing. The addition of facilities, equipment and labor are only a few of the considerations which business executives must discuss. Financial planners must also anticipate the finances which will be required to support employee empowerment programs, Total Quality Management (TQM), Just-In-Time (JIT), and ISO 9000 registration. All of these innovations are necessary to the achievement of world class operations.

The investment plan must also take into account the organization's strategic plan since, as we discussed earlier, the strategic plan is the deployment of company resources to achieve its vision for the future. The cost of taking the company into the future and the finances to make it happen are all part of the management decision-making process. As you can see from the World Class Process chart (Figure 6-3), there is a great deal of activity which

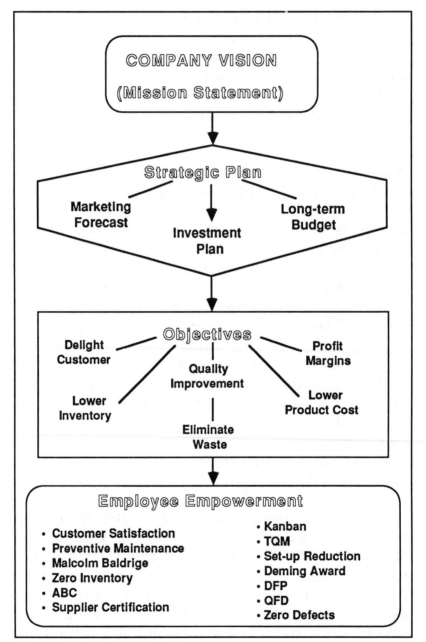

Figure 6-3. World Class Process

must be undertaken, from developing your organization's vision to planning investments to setting objectives to empowering action teams to achieve the desired results. The entire organization must be involved.

In order for the investment plan to be effective and realistic, the projected costs must be as predictable as possible. Projects which get identified in the strategic plan must be costed on a basis that includes all of the activities associated with the project. It becomes the responsibility of the project teams to project cost and savings comparisons as we have outlined earlier in this chapter. The emphasis should always be on generating savings for the entire organization. Let's take a look at an example which shows just how a world class improvement effort is put into effect. The EDI (Electronic Data Interchange) or bar coding project we have selected is a popular one at many companies today and demonstrates how a company should plan and anticipate the investment requirements needed to be successful.

Bar Coding Project

Let's imagine that your company is going to implement bar coding in the next year. A traditional approach would be to plan the investment based on the upfront cost of the bar coding equipment required to collect data automatically. This estimate is usually in the area of $25,000 and the project quite often dies here. A more effective method of ensuring that a project such as Bar Coding gets approved is to perform an activity based assessment. This assessment would determine the short-term costs and develop a plan to invest the savings which will be incurred into a long-term investment. The assessment can also help you determine what area of the facility will bring the most return for the least amount of investment. But to accomplish this, you will need to consider all costs associated with those current activities which will be eliminated once bar coding is implemented as well as the costs of all activities and equipment which will be needed by the project.

For example, let's look at the cost of implementing bar coding in the inventory area, starting with the receiving function:

Equipment and activity cost

Wand	$ 525 (with cables)
Wedge	330
Printer	2,000 (dot matrix)
Software	300 (to print labels)
Programming	528 ($22/hr **X** 24 hrs)
Training	132 (3 people **X** $22 **X** 2 hrs)
Instructor	100 ($50/hr **X** 2 hrs)
Accessories	1,000 (paper, labels, ribbon)
Total	$4,915

The project team should ask for $6,000 to get started. Now, let's see what the savings will be for a one-month period:

Savings (one month)

Labor savings	$1,320 (60hr/month **X** $22)
Data entry time	2,160 (40hr/month **X** $54 computer time)
Accessories	20 (forms, files, space)
Total	$3,500

Labor savings is associated with matching packing slips, filing, recording manual logs, etc.

Data entry time includes computer time to enter receipts, make corrections to receiving records and make inventory adjustments.

Based on this activity-based data, the team is well justified in asking for an investment of $6,000 since the company can expect a return on their investment in two months. The cost vs. benefits chart (see Figure 6-4) for this particular set of actions shows the course of the investments and payoffs. Not all projects, of course, will provide a payback this quickly. There are many factors which will influence your particular project.

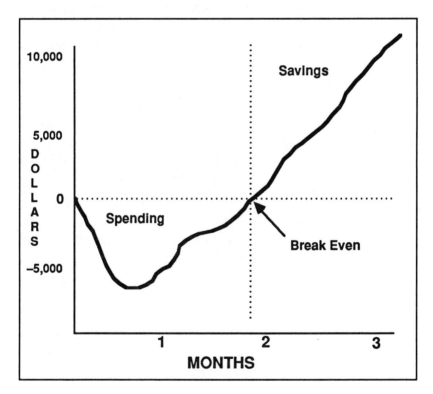

Figure 6-4. Cost vs. Benefit for Bar Coding Project

The savings in Figure 6-4 are real dollars. The chart shows that in less than two months the project will break even and that by the end of two months, there will already be a small savings which will increase over the ensuing months. These savings should be invested in the implementation of still more bar coding projects as a continuous improvement process.

Several points should be emphasized in our example. If the assessment you perform does not reflect potential savings, then your team and your organization's management should reconsider implementing the project. Also, if the initial investment is so expensive that you cannot expect a return in the short term (three to six months), then your team is probably trying to eat an elephant in one bite. Consider breaking the problem down into more manageable pieces which can be easily "chewed" and "digested." Our final point is to make sure that you have shown management how your cost vs. benefits assessment was performed and what the data means. This is necessary in order to get their commitment to invest in the first place and then to reinvest the savings into the continuation of the project. The World Class company creates an environment that challenges management to invest more money internally by investing in their employees' abilities to change the bottom line. The next chapter explores some of those investment strategies in a world of technological change.

Chapter Seven

PREPARING COST ACCOUNTING FOR TECHNOLOGY CHANGES

All over the world, companies are seeking methods for building products and providing services to their customers with increased speed, at lower costs and with guaranteed quality. Total cycle time and time-based management programs of one sort or another are evident in almost every organization. Everywhere we go, people are stressing the importance of benchmarking and mapping the process to gain improvement. Numerous companies, in fact, feel that pursuing this course is mandatory if they hope to survive.

Why, then, are companies not more successful in their approach to improving profits and reducing cycle time? You would think that, with all that activity, companies would be meeting their goals at an increasingly faster pace. The success rate of companies implementing new improvement programs is extremely low because in their race to benchmark, they fail to understand how to use the data they have gathered. As a result, most organizations are not in a position to challenge, let alone dominate, the world marketplace.

In our education programs held to help clients around the world, people are quick to point out that World Class improvement efforts are not working as well as they had hoped. The major culprit is a combination of finger pointing and unclear ownership of the process. This is what we hear from professionals throughout an organization:

- **Management won't commit the dollars or resources necessary to do the job right.**

- **People will not accept change.**

- **The union doesn't trust the program.**

- **Our product (service) doesn't lend itself to improvement.**

- **We don't have the time.**

To that last statement, our reply is a simple one: You will have plenty of time after your competition succeeds in implementing the program.

Recently, Mel got into a discussion with a participant at one of our Activity Based Costing (ABC) seminars. The woman was complaining that her management would not allow her to implement a Kanban project.

"Why should they let you implement a Kanban project?" he responded.

Even if management has the most advanced, World Class mindset and is looking for improvement ideas any place they can find them, they need to be sold on why they should implement any improvement effort. Merely to tell management that a program will improve this or eliminate that is simply not a good enough reason any more. Many efforts to introduce technology enhancements are meeting with little or no support, not because management is resistant to change, but because the introducers have failed to provide alternatives and convince management how the change will benefit the organization's performance.

Mel told the seminar participant that she needed to gain management's support with a costs vs. benefits analysis. Mel's favorite expression has always been: "In God we trust. All others must bring data." Since none of us is the Supreme Being, we need to bring along data and alternatives in order to convince management of the need to implement a new technology or philosophy. From our experience, that means figuring out the return on investment. The key to management commitment is answering the question of how does it affect the bottom line. Quality improvements and cycle time savings must be defined in terms of dollars. Executives will get real interested when savings are mentioned. Mel's recommendation, then, to the woman at our seminar was to prepare a costs vs. benefits analysis which would include the following:

- **Solid evidence of opportunities for improvement.**

- **Strategy for long-term/short-term improvements.**

- **How much it will cost to correct issues.**

- **Profile of potential savings.**

Cost accountants can assist improvement teams in the preparation of such analyses. They can help determine what the savings will be when a new technology is implemented in an organization. This means Financial personnel must be trained to understand new technologies and how costs will be affected by their implementations. In this chapter, we will focus on several improvement projects being considered for implementation by companies. We will be looking at what benefits organizations can expect. We are doing this because we feel it is vitally important that cost management understands and encourages technology improvements. Their specific role is making visible to both management and improvement teams the cost information which reflects the value and nonvalue-added activities of the area under consideration. These are the places which afford opportunities for improved future performance.

Group Technology
and the Savings of Cellular Operations

Manufacturing cycle time improvements are a challenge and an opportunity to any organization. Improvement should start with a process mapping of the plant and its operational flow. We advise companies to educate their improvement teams in how to prepare an operational map. A typical process flow diagram looks like the one in Figure 7-1.

OPERATIONAL MAP

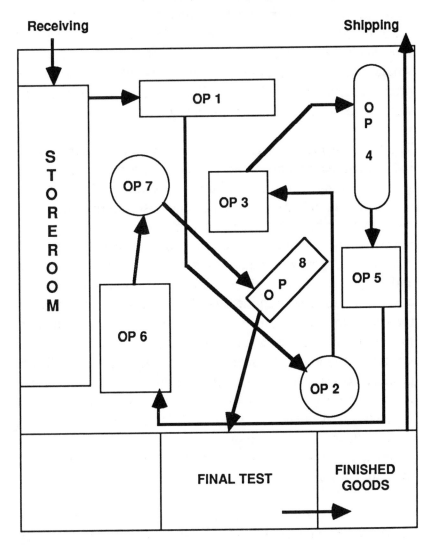

Figure 7-1. Operational Map.

The next step is to analyze each operation. The improvement team should be considering the time, resources and costs involved in completing the activities required to produce value-added activities. In addition, the team should study the relationship between value-added work performed as compared to nonvalue-added activities such as decision making, product movement and wait time. With the assistance of the cost accounting function, the team would take the information they had gathered and cost out the process as it is currently followed. Figure 7-2 shows how this chart would look for the flow diagram in Figure 7-1.

The costed resource chart can be very revealing. It can quickly indicate where a team should expend its efforts in eliminating waste, reducing cycle time and cutting product cost. It can also be very useful in helping Finance develop cost improvement standards that can be used in costs vs. benefits analyses. These cost improvement standards should reflect the parameters needed to cost the time spent working, waiting, decision making, setting up and staging a task as well as machine, equipment and facility costs. In addition, accounting should also encourage the improvement team to consider the reported percentages carefully in order to identify value-added and nonvalue-added costs. The Japanese stress that all motion, or movement of material, subassemblies or products, is waste and thus a nonvalue-added cost. Another task for the improvement team is to analyze why inventory levels exist in queue, movement and staging. This is material not being worked on and should be assigned a cost. Through education and training, it is possible to teach your employees how to calculate inventory costs.

The focus of costing resources is to recognize opportunities for improvement and then to implement the plan of action necessary to eliminate or reduce the waste activity. The primary job of the

COSTED RESOURCE CHART — CURRENT

OPERATION 570
COST & TIME ANALYSIS

Activity	Effort			Elapsed Time	I.C.C.	Machine & Equipment		Labor		Total Cost
	Work	Motion	Wait			Time	Cost	Time	Cost	
Kitting	50%	20%	30%	2.5 hr				2 hr	$ 37.50	$ 37.50
Staging	---	---	100%	8 hr	$27.60					$ 27.60
Move	80%	20%	---	15 min				15 min	$ 4.69	$ 4.69
Queue	---	---	20%	8 hr	$27.60					$ 27.60
Set-Up	20%	60%	100%	4 hr	$13.80			4 hr	$ 75.00	$ 88.80
Run	65%	25%	10%	4 hr		4 hr	$600	4 hr	$ 75.00	$675.00
Wait	---	---	100%	8 hr	$27.60					$ 27.60
				34.75 hr	$96.60	4 hr	$600	10.25 hr	$192.19	$888.79

Labor rate....... $18.75
Material Cost
  Issued... $3,450.00
Inventory
  Carrying Cost .008%/day
Machine Time... $150 hr

Figure 7-2. Costed Resource Chart (Before).

improvement team is to challenge every inefficiency on the chart and to search for the true causes. Part of this brainstorming and problem-solving activity requires the team to familiarize itself with new technology which can help their company not only solve the problems, but become world leaders in the product or service they provide. When improvement teams are exposed to the proper training and given the time to put their analysis tools to work, the results are sometimes astounding. Organizations benefit greatly with higher profits and much more involved employees. Let's take a look at what the improvement team at one company did with the process flow diagram in figure 7-1 after they were exposed to the technology of cellular operations. Figure 7-3 shows the new layout.

CELLULAR MAP

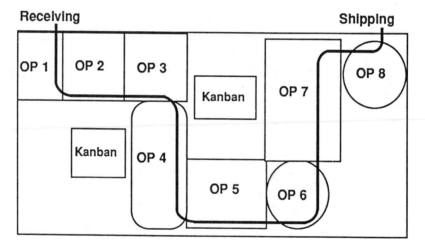

Figure 7-3. Cellular Map.

As you can see, the team made a number of improvements over the previous flow of activities. First of all, material requirements for scheduled production was delivered by suppliers directly to the floor in either of two Kanban locations. These locations were available to all operations requiring the materials and their use eliminated the need for material to be stocked, kitted and staged in an inventory storeroom as it was before.

Conveyor systems were also installed to move the product from one operation to the next, thus eliminating material between operations and improving cycle time. The operations themselves were laid out so that the flow of work from one operation to the next allowed the operator visibility throughout the process. Bar code scanners were also mounted next to the conveyors to automate the collection of information, move products, issue material and monitor performance.

The final test area was completely eliminated since operators were all trained in Statistical Process Control (SPC) and Six Sigma applications. Quality was now assured at each operation. In the new cellular plan, completed products were sent directly to shipping for delivery to the customer which also eliminated the need for finished good stores.

All of this sounds great, but what management requires is how much does the improvement save the organization in dollars and how much does it cost to implement. To answer these questions, the team should prepare a costed resource chart (see Figure 7-4) for the cellular process to compare with Figure 7-2.

Working with the two charts, the team can identify and report savings to management. Throughout the entire problem-solving and implementation phase, the team should be capturing all costs associated with improvement as well as documenting their steps. By that, we mean that costs should be captured for the time and

Activity Based Costing

COSTED RESOURCE CHART — Cellular Map

OPERATION 570
COST & TIME ANALYSIS

Activity	Effort			Elapsed Time	I.C.C.	Machine & Equipment		Labor		Total Cost
	Work	Motion	Wait			Time	Cost	Time	Cost	
Set-Up	90%	10%	---	30 min	---	---		30 min	$9.38	$9.38
Run	90%	10%	---	4 hr	---	4 hr	$600	4 hr	$75.00	$675.00
				4.5 hr		4 hr	$600	4.5 hr	$84.38	$684.38

Eliminated
• Kitting
• Staging
• Move
• Queue
• Wait

Reduced
• Set-Up
• Motion

Labor rate...... $18.75
Material Cost
   Issued... $3,450.00
Inventory
   Carrying Cost .008%/day
   Machine Time..... $150 hr

Figure 7-4. Costed Resource Chart (After).

material that went into procuring capital investments such as handling systems and bar code applications. Time should also include the number of hours that labor spent in meetings, attended training and education seminars and implemented the solutions. Labor, in this case, is defined as anyone involved on a team or doing work to support a team.

Even when companies are shown the potential savings which can result from the implementation of group technology, many still remain reluctant to change. We hear the same excuses over and over. They all seem to boil down to one statement: We're different! That's a difficult attitude to change, but every attempt should be made to educate and train your people to analyze your operations to determine areas of opportunity to reduce cycle time, improve customer satisfaction and eliminate nonvalue-added activities. This will result in a total cost reduction and make your organization more competitive.

Kanban Provides Momentum
for Inventory Reduction and Accuracy

The best translation of the word "Kanban" that I have heard is "easy to read signals." It accurately sums up the purpose of Kanban, which is to signal a need to the organization. That need can communicate the necessity for one of the following activities: Produce, replenish, plan or eliminate. In fact, Peter Grieco refers to Kanban in his book *Made in America: The Total Business Concept*, as a means to communicate. Kanban can indeed communicate these signals throughout an organization — administration, support, materials, manufacturing and service. The benefits of Kanban to organizations have been amply demonstrated and we refer you to *Made in America* for a more detailed discussion.

Kanban can come in many forms. We have seen it as a card, a board, a colored light, a box, a bin and even a shelf. The most popular form in America is the Kanban card as shown in Figure 7-5.

Pro-Tech **KANBAN** Card # 5 of 10

If found, please return to Professionals for Technology, Inc.

PART NO. ______________________

DESCRIPTION ______________________

SUPPLIER ______________________
Internal/External

LOCATION ______________________
Stock/WIP/FG

QUANTITY ____________ **U of M** ______

Part	Location	Quantity	Card #

Figure 7-5. Kanban Card.

A Kanban card is used to signal a demand to trigger a pull internally through various operations and to signal a demand externally for replenishment. A pull system starts with demand from customers. Sales then signals finished goods with a Kanban card to release the product so that it can be shipped to satisfy the customer's demand. This sets the stage for the issuing of other Kanban cards that signal the process to produce. This combination of signal followed by replenishment continues throughout all of the operations and eventually to the company's suppliers who then replenish the consumed parts, services or raw material. Figure 7-6 shows a typical Kanban pull system.

Cost accounting should look upon Kanban systems as an excellent opportunity to cut administrative costs. If implemented and used properly, Kanban can eliminate most paper and manual transactions which add cost to the purchasing process. Kanban cards which make their way through operations eventually find their way to Purchasing to signal a reorder. All Purchasing needs to do with the card is to scan the bar code in order to create a release against a contract previously established with the supplier. The card is then put into an envelope and mailed to the supplier instead of a purchase order or, in the case of more sophisticated systems, an electronic data interchange (EDI) is performed. Upon receipt of the card, the supplier responds by shipping the appropriate part in the standard quantity printed on the card. The card is returned with the shipment of parts.

Back at the company, the card is scanned by receiving to link the receipt of the parts with the contract release. Parts and card should then be delivered to the location within the plant identified on the card, whether that be inventory, work-in-process or any other location requesting parts. As you can see, a Kanban pull system is a closed loop that links together internal and external facets of

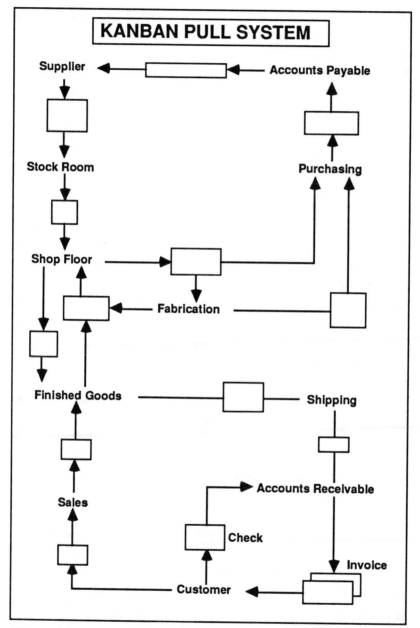

Figure 7-6. Kanban Pull System.

sourcing. The closed loop methodology accounts for Kanban's accuracy and savings. Let's now take a look at what Kanban has saved the company in the explanation that we have given.

SAVINGS FROM KANBAN

- No requisitions
- No order entry
- No purchase order
- No packing slip
- No receiving ticket
- No receipt transaction
- No stocking ticket
- No inventory update
- No invoice

Just in this short list, we have eliminated six forms and three manual transactions. Scanning the bar code on the cards sends automatic updates to purchasing, inventory and financial files. One of our clients, J.I. Case, implemented a Kanban system for eight part numbers at only one location on their plant floor. Three months after the implementation, the savings were recorded as $69,000! And that was just a pilot program! Management had no problem authorizing the continuation of the Kanban project into other locations.

Set-Up Reduction
and its Effect on Cost Savings

Set-up reduction programs have played a big role in helping companies to reduce not only the time to set up equipment, but lot sizes and costs as well. In fact, accounting should insist that improvement teams pursue more than just time savings. The perception that set-up reduction programs are implemented so that companies will have more time to produce product is only partially true. The real benefits come from cutting lot sizes to one, which allows for more flexibility in meeting customer demands for smaller quantities and which reduces cycle time. The agile manufacturing we discussed in the previous chapter places an emphasis on this high standard of performance.

The trend in business today is for customers to order smaller quantities in order to meet the Just-In-Time requirements of continuous improvement programs. Customers also want these lower quantities at the same unit price. This can create a problem for a company which is not flexible enough to produce smaller lot sizes. When that customer wants a quantity of 100 parts a week and the supplier keeps trying to sell them lots of 500 or agrees to lots of 100 but at twice the price, it won't be long before the customer begins to look for new sources.

When customers ask why we are charging double, a company's usual response is that they need to cover their set-up costs. But, what if there was a way to reduce set-up time? What would be its effect on cost? Should we implement a set-up reduction program? In our book, *Set-Up Reduction: Saving Dollars with Common Sense*, we begin by discussing the key motivations for addressing the issue of set-up reduction.

For this type of program to be effective on the cost side, accounting needs to determine what one minute of set-up costs so that the savings can be calculated for every minute that the set-up is reduced. This calculation needs to go beyond your current set-up standards. Companies will need to put an emphasis on uptime and on the savings which can be generated from being flexible enough to produce only what the customer wants. Zero set-ups and equipment which is immediately available for use are absolutely vital in the agile manufacturing environment. Producing on demand, needless to say, also has a large effect on inventory reduction and the savings generated there must go into your calculations as well. Inventory carrying costs are, in fact, an excellent way to show the benefits and savings which accrue from building to customer demand as Figure 7-7 shows.

INVENTORY CARRYING COSTS AS SET-UP TIME IS REDUCED

Set-Up	Lot Size	Production Cost Per Piece	Daily I.C.C.	Daily Cost
4 hr	5,000	$48.00	.06%	$144
2 hr	2,500	$48.00	.06%	$ 72
1 hr	1,250	$48.00	.06%	$ 36
30 min	625	$48.00	.06%	$ 18

Figure 7-7. Inventory Carrying Costs.

As Figure 7-7 points out, the more set-up time is reduced, the less inventory carrying costs you incur. And, obviously, the lower your inventory costs, the less you need to charge for the smaller lots which many customers demand. As a consequence of the lower prices and more flexible amounts, you stand to also gain more business. The effect of set-up reduction on costs works in this way. Should a customer desire 500 pieces and your set-up is 4 hours, producing lots of 5,000 would make it necessary for you to absorb 12% a month on the excess 4,500 pieces. That amounts to $25,920 in carrying costs alone. If you could do the same set-up in 30 minutes, the inventory carrying costs on the excess of 125 pieces would only be $720 a month. Thus, in order for you to sell 500 pieces when your production runs are 5,000, you would need to charge your customer more than twice the cost per piece in order to recover inventory carrying costs.

This is just another example of how Activity Based Costing reveals hidden costs which affect the profitability of your production process. Set-up reduction is mandatory in a technology hungry market. You must encourage your organization to start the process of reducing set-ups in order to remain competitive. As the example above shows, the cost savings are ample evidence of why this is necessary.

Data Collection Advances —
Providing More Information with Less Effort

Many companies seem to be avoiding the issue of collecting and reporting more realistic information because they believe it will require additional effort and cost on their part. And yet, many of our clients tell us how unhappy they are with the information on labor, inventory accuracy and shop floor status that their current systems collect.

We have found the biggest concern to be inventory accuracy. You would think that auditors would insist on inventory being accurate given the emphasis they place on how inventory is valued. The truth is, however, that inventory accuracy is not that important to accountants. We have been through enough annual physical inventories to know that as long as you count it and have numbers that can be valued, the accuracy of the counts is not critical. We are sure that there are some auditors who would take exception to our assessment, but we would maintain that the level of accuracy is not sufficient for the manufacturing environments of today which exist in a Just-in-Time, total quality and agile marketplace. What we need to find in the new business world is a system of monitoring the ins and outs of inventory more accurately without adding effort or cost.

Fortunately, the new business environment has brought about technological advances which can collect more accurate data with which to make business decisions. And, the new technologies can do so without extra effort or cost. Our discussion earlier in this chapter about bar coding Kanban cards certainly demonstrates how data collection in the materials area can be made easier.

The shop floor has also benefited from more advanced methods of collecting data. The emphasis is on eliminating the need for people entering data on work sheets or into terminals. For example, quality results have been expanded to include not only the number of defect, but the reasons why the product was defective. Menu boards with different bar codes for each kind of defect are scanned by the operators as defects occur. The collection of this valuable data allows operators to analyze why the defects are occurring and to suggest solutions based on far more accurate data.

The collection of labor data has always been a difficult area to control when operators have been asked to manually input hours and rates on time sheets or cards. With bar coded badges, work papers and menu boards, however, the data just needs to be scanned in order to capture the labor information more accurately. Pro-Tech suggests going even further. We ask why does labor need to be captured at all in the traditional manner in agile manufacturing environments. Labor in many companies accounts for less than 5% of the product cost. It's time to put our focus on the supply and customer side.

Cost Accounting should always be looking to challenge traditional methods of data collection in an effort to find easier, more accurate and less costly means. Automation of data collection through the use of data collection techniques is not that expensive and the returns on investment are high and realized very quickly.

One last area that Cost Accounting should consider is Electronic Data Interchange (EDI). Companies who have automated data collection internally are so pleased with the outcome that they have wanted, as a next step, to be able to communicate just as accurately and quickly with their suppliers and customers, that is, with external entities. EDI systems allow for this type of exchange of information. For example, customers can send their demands to your plant electronically. In turn, you can electronically transmit your own demands to your suppliers. All this can be executed without costly paper through the mail activities. And, with EDI, there are far fewer disputes between customers and suppliers since the accuracy of the communicated demands is far better than paper methods. Who knows? Maybe we can have our customers get components released directly from our suppliers in the future.

The clear fact is that technology advances every year whether you choose to get involved or not. But if you don't, you can be fairly certain that one of your competitors is looking into or implementing the new technology to gain an advantage in the marketplace.

Supplier Management and the Cost/Price Comparison

The purchasing of materials and raw materials from a supply base has long been an area of contention in the manufacturing environment. Customers want small lots free of defects delivered on time at the lowest possible cost. Suppliers want guaranteed forecasts of large lot deliveries at prices that cover their costs and they want their customers to pay on time. From this situation, we have seen some very aggressive negotiations that end up promoting a less than cooperative relationship. Many companies, however, are starting to realized that this approach to the procurement of material is counterproductive for both sides. They are beginning to use a different approach in which a win/win relationship for both the customer and the supplier. This new partnership approach is promoted by the methodologies of cooperation introduced by Supplier Certification programs.

Above all else, Supplier Certification programs allow for clearer communication between both partners. The focus of the relationship is on quality, delivery, quantity and profit to be shared by all. In his book, *Supplier Certification II: A Handbook for Achieving Excellence through Continuous Improvement* and *Just-In-Time Purchasing: In Pursuit of Excellence*, Peter Grieco points out that both customers and suppliers should be actively involved in cost control programs and should demonstrate and share results. Some of the cost control activities suggested in his book are listed below:

- Customer and supplier should be able to demonstrate that they are actively involved in programs to control and reduce costs in areas of waste reduction, supplier cost control, productivity improvement, efficiency improvement, and technological advancement.

- Customer and supplier should work together to set cost standards against which operations can be meaningfully and usefully measured. These standards should mirror the usage, efficiency, productivity and other quality and control measures utilized in the production process.

- Customer and supplier should determine to what extent standard and other costs are allocated appropriately to the supplier's product which is purchased.

- The supplier's Accounting staff should interact with its Operating staff in the reporting of frequent and useful results.

- Customer and supplier should both be familiar with the cost of quality.

- Customer and supplier should control lead-time costs.

- Customer and supplier should effectively control raw material, intermediate material and finished goods inventories.

- Customer and supplier should share cost information in order to provide mutual benefits through a commitment to long-term supplier relationships which protect supplier profitability while assuring the customer of acceptable costs.

- Customer and supplier should be willing and able to provide the components of cost.

The bottom line in a customer/supplier partnership is to cut the cost of purchasing materials required to manufacture products. But Cost Management has to encourage Purchasing to look beyond the price being paid. There are costs associated with many other purchasing activities that are just as expensive as the list below indicates:

- **Cost of preparing a purchase order (P.O.).**
- **Cost of paper.**
- **Cost of mailing.**
- **Cost of receiving.**
- **Cost of counting.**
- **Cost of inspection.**
- **Cost of Material Review Board (MRB).**
- **Cost of defects.**
- **Cost of disputes.**

In our book, *The World of Negotiations: Never Being a Loser*, we show that companies pay far above and beyond the listed price for material or parts. It is the job of Supplier Management teams to become aware of these additional costs and then to work externally and internally to bring them down. The teams should also work toward developing measurements that reflect the focus of the entire effort. For example, one of the goals of a Supplier Management program is to reduce the supply base. One measurement, then, should total the current supply base to arrive at a benchmark figure. The program would then work toward lowering that number. A cost associated with each supplier should also be computed. This cost may be as high as $5,000 for each supplier developed. The intent would be to give a numerical focus to the reduction efforts of the teams.

ABC must, however, calculate an accurate figure which is directly associated with the cost of maintaining a supplier at your company. Some of the cost elements which could be calculated are listed on the next page:

- Research and development engineering costs.

- Supplier selection, surveys and audits.

- Supplier visits.

- Computer file space used to store the supplier's history file, purchasing history, payment history, etc.

- Cost of forms such as purchase orders, receivers, vouchers and checks.

- Matching of invoices.

- Time spent on the computer entering purchase orders, receiving entries and voucher entries and time spent processing checks.

- Cost of acquisition.

- Inventory storage space for supplier's parts.

- Supply management.

- Human resource activities.

As teams work with suppliers in all of the areas mentioned, they begin to implement new technologies or cost-reduction programs, both partners will experience the benefits of a win/win relationship and cost reduction. Focus, however, should always be maintained on associating each activity or technology with cost savings.

Summary

In this chapter, we have addressed how technology changes can be costed for continuous improvement opportunities. The entire company needs to be educated about the real cost benefits of these improvements. And the teams which are developed to analyze and implement improvements need assistance in identifying real savings and in demonstrating costs versus benefits. Companies need to make a commitment to an improvement program which embraces, rather than avoids, technology changes.

Chapter Eight

LIFE CYCLE COSTING: From Development to Market Satisfaction

In the increasingly competitive marketplace of today and tomorrow, new technologies and methodologies are forcing organizations to become better prepared at cooperating internally and externally. This movement toward improvement is the driving force behind those companies seeking a position of prominence in the future. Improvement programs are bringing Engineering, Manufacturing, Sales, Quality and Financial functions together in an effort to "do it right the first time," whether that means the production of a product or the providing of a service, or both.

Whatever new technologies and methodologies are used, the targets for success listed below will remain the same and companies who understand them will succeed:

- **Quality will conform to customer requirements.**
- **On-time delivery will be expected.**
- **Sales will be awarded to the lowest cost provider.**

These targets will remain constant regardless of products, services, environmental factors or world conditions. But, new technologies and methodologies without cost controls can turn an organization's attempts into a disaster. Traditional practices encourage us to assign technology costs to the costing equation as period expenses. In doing so, however, technology costs lose the relationship they have to the products and services affected by their implementation. Design people may be content with this arrangement, but the financial community should not.

New technologies and methodologies must give rise to new methods of accounting for improvement. One such method is the concept of Life Cycle Costing.

DEFINITION

Life Cycle Costing is the accounting of cost in relationship to the activities performed in developing a product or service from creation through retirement.

As the definition demonstrates, Life Cycle Costing uses a functional approach in the gathering of cost data. Each area within an organization that plays a part in the product's or service's life cycle will be analyzed to determine its cost impact at the product level. The areas that will contribute to costs during the life cycle are the following:

LIFE CYCLE COSTING
<u>FUNCTIONAL AREAS</u>

- **Design/Engineering**
- **Planning/Scheduling**
- **Sales/Marketing**
- **Costing/Estimating**
- **Materials/Purchasing**
- **Receiving/Shipping**
- **Payables/Receivables**

This chapter's focus will be on assisting the cost accounting function to capture, plan and control life cycle costs. This will entail estimating, evaluating and improving upon the elements of cost from all functional areas within an organization. We will look at designing cost targets which attack and reduce nonvalue-added activities in an aggressive manner. Lastly, we will introduce a Life Cycle Model to help you facilitate the process of reporting results of budgeted and actual performance.

The Capturing, Planning and Controlling of Cost Data from all Functional Areas

The objective of Life Cycle Costing is to capture activity costs from all functional areas. These functional costs divide into two categories:

- **Development Costs.**
- **Operational Costs.**

Development costs are nonrecurring costs expended during the period when a product is conceived, designed, marketed and finally released to production. Companies that have successfully implemented Life Cycle Costing have learned that the earlier in a product's development that cost reduction is considered, then the easier it will be to control costs throughout the life cycle. Some have estimated that 90% of a product's total cost is designed into a product. Thus, the emphasis that Life Cycle Costing puts on designing out as much cost as possible during the development stage and before production begins. We believe that far too many companies are currently spending billions of dollars to reduce costs in the wrong area, that is, in operations. Recovering costs in the operations stage is very difficult since the costs have not been captured, planned and controlled from the outset of the product life cycle. Figure 8-1 shows just how large of an influence that design has on life cycle costs.

Traditional accounting practices fail to capture and report these influences. The associated costs are clearly visible at the shop floor level, but, in most companies, personnel there are powerless to effect any improvement. Activity Based Costing, however, can make these costs visible to the Design function which can effect improvements in current and future development efforts. Design has a unique window of opportunity (see Figure 8-2) to make total cost reduction a target of the development phase. Unfortunately, the elements of cost which present the biggest opportunity are usually not in the mindset of traditional design personnel. Their mindset has been shaped by pressure from management to expedite the product to market. What usually happens in this design environment is that corners get cut and unresolved issues are passed on to Production to overcome.

With the proper training and education, however, design person-

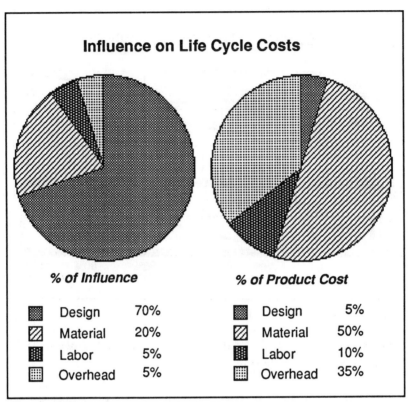

Figure 8-1.

nel can regulate their mindset so that they seek out target costs
such as those listed below.

DESIGN TARGET COSTS

Fewer Parts	**Simple Process**
Multifunctional Materials	**Simple Assembly**
Elimination of Hidden Costs	**Design Quality In**
Design Inspection Out	**Less Transactions**
Less Paperwork	**Standardized Cartons**
Less Transportation	

Total Life Cycle Savings

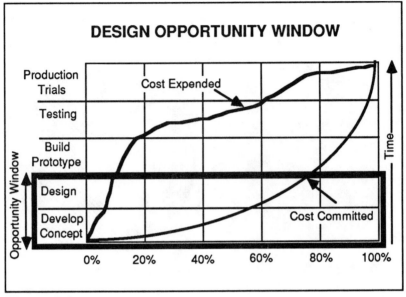

Figure 8-2.

Even though no one likes rules, the rule behind the target costs above is a very simple one:

DESIGN IT RIGHT EACH CRITICAL TIME

The direct rule not only gives Design the proper focus but helps to coordinate the strategies and goals of all the functions within a company. Furthermore, the rule not only allows for more creativity in product design, but in cost reduction as well.

Operational costs are recurring costs that occur with every production run. In Life Cycle Costing, the operations function is encouraged to work with design and development personnel in the identification of all nonvalue-added activities accumulating in the product's life cycle for eventual reduction or elimination. The feedback loop between the two functions makes visible the cost drivers behind the target goals of the design phase. Accounting's role in this exchange of information is to relate the nonvalue-added activities to the cost drivers they represent as shown below:

LIFE CYCLE COST DRIVERS	
Driver	**Cost**
Late to market	Expedite charges
No specifications	Idle labor, material, machines
Errors in specifications	Scrap and rework
Failure to communicate	Increased cycle cost
Unrealistic tolerances	Excess labor and material

Every effort must be made to eliminate or reduce these costs. However, not all problems can be predicted and so accounting must plan on some percentage of cost being expensed during the production phase. Likewise, most design cost issues are planned and captured as failure costs in the Cost of Quality report.

Activity Based Estimates that Evaluate
Total Product Cost and Profitability

The estimating function also undergoes an enhancement when Activity Based Costing is introduced into an organization. The ABC estimate, as would be expected, provides a much more realistic picture of what a product's or service's total cost will be since the estimate is based on the principle that costs are assigned according to an activity's use of resources. Indeed, the preparation of cost estimates using ABC categories helps an organization in the identification of improvement opportunities. The ABC estimator highlights reduction opportunities by identifying the causes of cost, or the cost drivers.

A cost driver is an activity or task which creates or influences a cost element. In making cost drivers visible, an organization can then find ways to plan and control costs by initiating actions which reduce or eliminate nonvalue-added activities. This sequence of events is particularly important in the design phase when an estimator has an opportunity to identify and plan alternatives. Cooperation between Design and Cost Estimating in the formation of a cost reduction team is instrumental in focusing on cost reduction opportunities which go beyond material and labor areas, as can be seen in the list below:

In order to maintain profit levels and competitive prices, we need to reduce costs and eliminate waste throughout the organization. Figure 8-3 shows some of the roots of product cost in which waste activities must be identified for improvement. This streamlining of our operations will require us to eliminate the problems associated with waste, such as:

- **Scrap.**
- **Supplier Delinquencies.**
- **Purchasing Lead Times.**
- **Change Orders — Purchasing and Engineering.**

- **Long Set-Ups.**
- **Machine Downtime.**
- **Equipment Imbalances.**
- **Inspection Backlogs.**
- **Paperwork Backlogs.**
- **Absenteeism.**

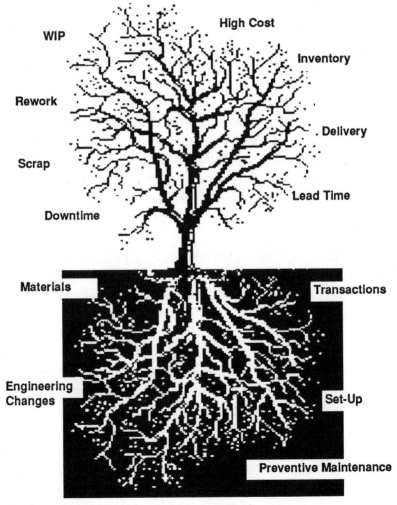

Figure 8-3.

In order to estimate the cost of waste, the estimator must first set a benchmark for each item. Then he or she should work with a team to determine what data should be collected, how it should be interpreted and how it influences the total cost. When these criteria have been established, information collection should start immediately in order to capture what the current costs are for each item. The estimator then keeps tabs on any activities which are trying to improve upon the benchmarked figures. This information is then computed into the estimate. For example, if a team or program is in place to reduce scrap by 20%, this needs to be reflected in the estimate.

For estimators who are working with a design team, there are two ways to create an estimate: 1) From scratch or 2) From an existing parts estimate. Whatever method is employed, the estimator's role on the team is not only to produce an estimate but to provide his or her expertise and knowledge in order to achieve the lowest cost possible.

One method for ensuring the lowest cost possible is to incorporate standardization into the design. The estimator should challenge every new part or process that is being designed with the following questions:

- **Why do we need this part?**
- **Will an existing part meet the requirement?**
- **What value does a new part add to the product vs. the cost of the new part?**

Our experience has been that an improvement in the design of parts often has a more powerful impact on total cost reduction than a process improvement. This is true, in part, because the standardization or elimination of parts has additional benefits that are more difficult to quantify. For example, standardization usually results in improved reliability, a reduction in inventory and lower production control costs.

What has become particularly clear to us is that companies can realize the greatest cost reductions when they begin the estimating process early in the life cycle. Estimates done in the design stages are proactive. They help the development team find the lowest total cost for the entire life cycle, not for just one particular phase. An estimate should be prepared in parallel with the design phase. If you wait until after the product has been designed, it is too late for significant cost reductions.

The biggest difference between traditional and advanced cost estimating methodology is in recognizing opportunity. Advanced methods focus on identifying cost improvement. Costly resources are identified and challenged in order for the organization to achieve the lowest cost products. Cost elements are looked at individually and questioned as to the value they add to the product. Some of the questions which must be asked are as follows:

MATERIAL COSTS

How many SKUs (Stock Keeping Units) are we buying?

Are we using the right components?

Do we purchase quality parts?

What is the lead time on delivery?

How good are our suppliers?

Do we count parts?

What is the cycle time?

Do we have to test parts?

Are suppliers certified?

MACHINES

Are the right machines being used?
Are machines being maintained?
Are operators trained?
Are operators given specifications?
Are statistical measurements used?
How long are changeovers?
Do machines have proper tooling?

HUMAN RESOURCE TASKS

How frequently are materials handled?
Who is performing testing/inspections?
How much paperwork is done?
How much manual work is done?
How much computer input is done?

OVERHEAD

How realistic is the overhead spread?
Is the cost of quality captured?
Is rework captured by product?
Are allocation methods specific enough?
Is total product cost available?

Establishing Target Cost Objectives

The organization which implements Life Cycle Costing must
incorporate advanced design methodologies in order to design

products within established target costs. This will require establishing competitive costs and challenging the design team to work with estimators and other functions to find the lowest total cost alternative. If a design does not fall within the target cost, then the team must make design changes until it does. This practice is called Design to Cost (DTC) and it is closely aligned with Design for Producibility (DFP). The dominant principle of both concepts is to reduce total cost. More specifically, the main advantages of advanced product design methodologies are:

Lower Costs — Labor, Material, Overhead

Reduced Tooling Costs

Fewer Number of Parts

Increased Flexibility in Manufacturing

Early Product Development

Quality Designed into the Product

Both advanced methodologies are based on economic principles which must become part of your corporate culture and company vision. Your organization, if it is a typical one, will need to address a number of issues which can hinder the adoption of advanced design and Life Cycle Costing. First, you will need to take resources allocated for traditional methods and put them into the more advanced practices. Second, you will need to increase the communication and cooperation between functions. And third, you will need to start educating and training your people in the concepts of DFP and DTC. Some of the economic principles which you will need to cover are listed on the next page:

ECONOMIC PRINCIPLES

- Reduce complexity.
- Standard materials and components.
- Standard design of the product itself.
- Fewer components.
- Determine tolerances.
- Use processible material.
- Communication and participation of operators.
- Group similar processes.
- Avoidance of secondary operations.
- Develop appropriate expected levels of production.
- Utilize special process characteristics.
- Avoid process restrictiveness.
- Testing, not inspection.
- Doing It Right the First Time.

Life Cycle Model for Planning and Budgeting a New Product

The reality of Life Cycle Costing is that most support functions within an organization know little about costing. The day-to-day pressures with which any organization deals leaves little time for cost considerations. Furthermore, in most business environments, personnel in the costing function are too busy with financial matters to develop life cycle costs. It's the old story of everybody being too busy, so nothing gets done.

One solution to this dilemma is to utilize the Life Cycle Modeling we have included here. The model you use should consider costs by function and activity as well as provide guidelines and formulas that assist in preparing life cycle costs. The formulas in your model will be specific to your company's functions and activities. Each formula will need to represent cost expenditures for labor, material, machine, time and waste.

LIFE CYCLE MODEL

DESIGN FUNCTION

Activities	Freq.	Formula	Life Cycle Cost
Develop concept			
Detail Design			
• Drawings			
• Bill of Material			
• Routing			
File Updates			
• Parts Master			
• Bill of Material			
• Routing			
Documentation			
• Work instructions			
• Test instructions			
• Packaging instructions			
Prototype			
• Set-Up			
• Runs			
• Validation			

DESIGN FUNCTION (continued)

Training Operations
- Documentation
- Set-Up
- Process
- Inspection
- Packaging

Trial Production
- Labor
- Machine
- Materials

Subtotal

MARKETING AND SALES

Activities	**Freq.**	**Formula**	**Life Cycle Cost**
Market Research			
Advertising			
Surveys			
• Market			
• Customer			
Market Forecast			
Sales Support			
• Travel			
• Phone and Fax			
• Administrative			
Order Entry			
• Computer			
• Postage			

Subtotal

MATERIALS

Activities	Freq.	Formula	Life Cycle Cost
Supplier Sourcing			
File Updates			
• Vendor File			
• FOB Terms, Ship Via			
Computer Time			
Supplier Follow-Up			
Receiving Support			
• Material Handling			
• Counting			
• Computer Time			
• Administrative			
Inspection			
Inventory			
• Stocking			
• Kitting			
• Cycle Counts			
• Computer Time			
Shipping			
• Kitting			
• Packaging			
• Materials			
— Shrinkwrap			
— Pallets			
— Labels			
— Boxes			
• Freight			
		Subtotal	

PRODUCTION

Activities	Freq.	Formula	Life Cycle Cost
Material Handling			
Documentation			
• Work Instructions			
• Test Instructions			
Set-Up			
Materials			
Labor Run			
Machine			
Labor Testing			
Rework			
Scrap			
Computer Time			
• Job Status			
• Labor Reporting			
• Quality Reporting			
		Subtotal	

Design Subtotal: _____

Marketing and Sales Subtotal: _____

Materials Subtotal: _____

Production Subtotal: _____

Grand Total: _____

Chapter Nine

COST MANAGEMENT:
Providing the Emphasis
for Continuous Improvement

An organization that identifies areas in need of improvement is not guaranteed success. The organization must move beyond identification and seek to motivate its people to improve. This is the only way to ensure long-term success. Companies have traditionally looked to incentive programs which offer monetary rewards as their chief means of soliciting and effecting improvements. However, improvements of this type look only at increases in productivity. Little emphasis is placed on cost savings. In fact, most employees are inadequately exposed to the cost issues that affect the bottom line. **More attention is paid to the production line than the bottom line.**

Cost management can change this orientation by providing the impetus to seek continuous improvement and by showing how to monitor and analyze costs. As we have discussed at length in this book, Activity Based Costing enables the organization to relate dollar amounts to both value and nonvalue-added activities. It is the activities which do not add value or which contribute to waste that afford the best opportunities for continuous improvement. The role of cost management is to make these cost figures visible to the appropriate people in order to motivate them toward cost improvement.

Determining Cost Responsibility and Accountability within the Organization

In order to be most effective, the results of Activity Based Costing need to be distributed throughout an organization and down to the levels which actually engage in the activities needing improvement. Distributing the results is a good start, but not enough. Responsibility for these results must also be established.

WHO IS RESPONSIBLE FOR...??

Administrative Cost
Design Cost
Engineering Cost
Procurement Cost
Manufacturing Cost
Distribution Cost

Traditionally, these categories of cost are budgeted and compared against actual performance. However, these categories as they are defined above are far too broad to allow for any serious improvement efforts. Each category needs to be broken down into cost elements that are descriptive of the activities which are actually performed. This is the backbone of Activity Based Costing. Cost management can best identify these cost elements and obtain their data by allowing each function (or department) to do the identification and data collection itself. By contributing to the identification and collection process, each function is more ready and comfortable with taking on the responsibility of effecting improvements in the identified cost elements. Each function should also be encouraged to capture its daily activities and cost management should get involved in helping to establish the set of cost elements. Figure 9-1 shows how the cost elements for an Accounts Payable function might look.

This list could continue depending on what activities each function performs. When compiling such a list, don't spend too much effort or time on collecting how much time is spent on each activity. All that is needed is a sampling to create a benchmark. Then it becomes the responsibility of the function to cost out these activities and begin the improvement process. In our Accounts Payable example, we would expect this function to challenge each of these activities with the following questions:

WHY???

Why perform the activity at all?
Why does the activity take so long?
Why does the activity cost so much?

ACCOUNTS PAYABLE	
ACTIVITIES	**COST ELEMENTS**
• Separating and receiving invoices	Labor
• Invoice error activities	Labor
• Voucher entry	Labor, computer time
• Check processing	Labor, materials and computer time
• Stuffing envelopes	Labor, materials
• Mailing checks	Labor, postage
• File maintenance	Labor, computer time
• Error processing	Labor, computer time
• Meetings	Labor
• Communications (Phone and Fax)	Labor, communication charges

Figure 9-1.

Improvement will come when these questions are answered using problem-solving techniques which provide solutions that eliminate, reduce or automate the activity. Relating costs to these activities and then charting the process of their reduction is a powerful incentive to improve.

Although responsibility for cost improvement may be assigned to specific levels of the organization, whether or not they respond depends on if they are held accountable. Accountability for any cost improvement activity lies with the next level up in the organization. That level should establish a target cost that encourages improvement once they understand the opportunities for cost reduction which the responsible level has uncovered. Both the responsible and accountable levels should communicate often on how the improvement process is proceeding.

Distinguishing between Cost Improvement and Cost Savings

It has been our experience that many improvement efforts are not given the praise or attention they deserve because the savings were not immediately visible on or to the bottom line. This type of mindset, in which savings are only seen in terms of reduced labor or material, is damaging to the concept of continuous improvement. Some organizations even believe that there are no savings in labor unless people are let go. But this attitude has to change. Cost management must be willing to accept that any effort which contributes to the list below is going to eventually affect the bottom line favorably.

IMPROVEMENT EFFORTS WHICH EVENTUALLY AFFECT THE BOTTOM LINE

Quicker
Easier
Safer
Better

Any effort to make a product or service quicker, easier, safer or better will almost always have a highly favorable influence on other areas in an organization. Customers, for example, want their products or services delivered or provided more quickly. The result will often be increased sales. Activities which are made easier often eliminate other activities, such as error collections. Any improvement that makes an activity safer should certainly be recognized as a cost savings. As for better, it can be defined in many ways — quality, job satisfaction, customer satisfaction, etc. Eventually, cost savings will accrue from any effort made in these four directions.

All savings need to be encouraged in an organization. Many companies that are involved in continuous improvement programs have established various methods for participants to collect improvement savings and cost savings. One of our clients in England provides its improvement teams with a cost control sheet which helps team members assign costs and savings. Figure 9-2 shows what such a sheet looks like.

As each team prepares measurements of their activities, the sheet becomes a tool of how much cost they have incurred to get their savings. All costs must be captured in order to show costs versus benefits. Teams as well as individuals involved in cost improvements should prepare these charts to demonstrate the results of their efforts. Cost management should work with teams to develop cost measurements and improvement parameters in addition to charts which show cost savings to the bottom line. Figures 9-3 and 9-4 show examples of a Cost Improvement chart and a Cost Savings chart.

Cost improvement appears to be higher than actual cost savings but that is because cost improvement is measured out into the future whereas cost savings only reflect to-date savings. For example, a waste reduction that results in a savings of $1250 per month will show up as a cost improvement of $15,000 which

COST SAVINGS CONTROL

CATE-GORY	DESCRIPTION	APPROVED BY	COST TO	TEAM RECORD? Cost	TEAM RECORD? Savings	COMMENTS/ GUIDANCE
A	Maintenance needs identified during study which one would expect to be normally carried out: Stiff/slow adjusters, worn adjusters, badly fitting guards due to damage — slowing access, etc.	Engineering	M & R	No	Yes	When problem identified, then normal maintenance procedures should be followed — team member from Eng. or Mgmt to act
B	Any item falling into Cat. A but costing less than £20: Purchase of fasteners, minor modifications, etc.	Team	M & R	Yes	Yes	Team member from Eng. or Mgmt. can act
C	Any item not falling into Cat. A or B but costing less than £300	Team	RFJ	Yes	Yes	Team identifies, proves, costs and approves work
D	Any item not falling into Cat. A or B or C but costing more than £300	Steering Committee	As Appropriate	Yes	Yes	Team identifies, proves, costs and formulates written justification; submits to steering comm.

Figure 9-2.

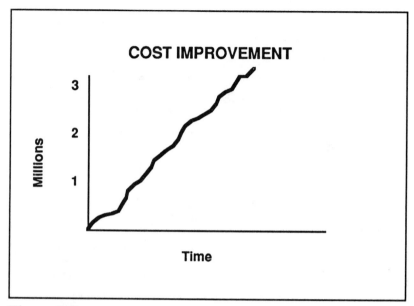

Figure 9-3.

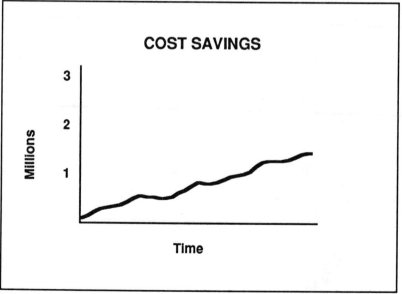

Figure 9-4.

reflects 12 months of savings. We encourage cost management to maintain both measurements and the team should use the cost improvement chart. The cost improvement chart is used because it helps to motivate teams to drive figures higher.

Providing Visibility into the Continuous Improvement Environment

The objective of capturing and reporting measurements is to provide visibility into the accomplishments of the continuous improvement process. It has been our experience that teams which take on the responsibility of reporting improvement activities also take on ownership of the results. It is important that these measurements be as accurate and easy to collect and maintain as possible. Teams of employees can provide management with pertinent measurements when the teams are given direction in how to do so.

A method which we have found to be very effective is to assign one member of the team the responsibility for collecting all of the team's measurements. Similarly, an employee from cost accounting or a member of the steering committee should be assigned the task of collecting all of the team measurements and providing a summary of the entire improvement process.

The intent of training team members in what constitutes a measurement, what is a cost and what is a saving is to improve the accuracy of team results. We encourage organizations to let their improvement teams maintain as many meaningful measurements as they desire. At the very least, every team should report costs versus benefits. Cost management can aid the process by developing cost criteria baselines such as the following for the teams to use:

COST CRITERIA	$ VALUE
One minute of set-up	$.40 a minute
One day of cycle time	$1,700
One minute of downtime	$80
One minute of idle time	$22
Daily component and product scrap value	Actual dollars
Stockout cost	$250
Material handling charges	Labor rate **X** time
Computer data entry cost	$2.44 a minute
Lost business cost	Product or service value

Each of these baseline criterion will provide the team with visibility into a number of opportunities which members can pursue. Furthermore, teams need only collect data on time to calculate measurements for each of the criteria.

Let's turn our attention now to some measurements that a set-up reduction team uses at a client of Pro-Tech's in England. The team created a summary sheet (Figure 9-5) and a cost versus benefits graph (Figure 9-6) to demonstrate their progress. The key to the measurements they took was how much one minute of set-up cost their company. The cost (all figures for this example are in American dollars) was calculated by cost management to be $0.47 per minute. This figure represented both labor time and fringe benefits.

Since the team spent $479.00 to save $1,123.77, they showed a net savings to the company of $644.77 per set-up. If, however, this

SET-UP REDUCTION SUMMARY SHEET

SEGMENT	Current Time/Min	Target Time/Min	Achieved Time/Min	Time Save/Min	Dollar Savings	Cumulative Savings
1) Crimping head replacement	175	90	38	137	$ 64.39	$ 64.39
2) Crimping head set-up	864	400	25	839	394.33	458.72
3) Green belt adjustment	247	120	13	234	109.98	568.70
4) Foil length adjustment	1010	500	165	845	397.15	965.85
5) Set up foil reel	383	200	47	336	157.92	1123.77
TOTALS	2679	1310	288	2391	$1123.77	

Time in minutes
Std. Cost = $0.47 per min

Figure 9-5.

improvement was costed, the measurement would look a bit different. Look at the chart below:

Set-up savings	**$ 1,123.77 per set-up**
5 Set-ups per month	**$ 5,618.85 per month**
12 Months	**$67,426 per year**

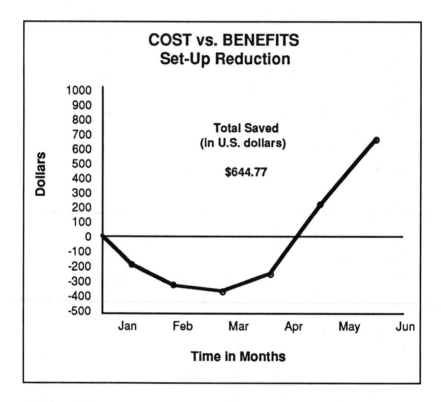

Figure 9-6.

The team could also calculate savings for inventory reductions which were the result of eliminating the need for a larger lot size per run. These savings are depicted in Figure 9-7.

Set-Up Time	Lot Size	Percent Reduced	Cost		
			Material	Carrying	Total
2679 minutes	100 pallets	0%	79,500	9,937	89,437
1310 minutes	50 pallets	50%	39,750	4,968	44,718
670 minutes	25 pallets	75%	19,875	2,484	22,359
288 minutes	12.5 pallets	10.75%	9,937	1,242	11,179

INVENTORY REDUCTION

Material is $795 a pallet
Inventory Carrying Cost = .125% per day

Figure 9-7.

Conclusion

As we have demonstrated, a team which is focused on savings can generate very favorable results when it is provided with the proper parameters to measure. In our book, *World Class: Measuring Its Achievement* (PT Publications, Palm Beach Gardens, FL), we provide our readers with a foundation upon which they can develop their own baselines. Carl Cooper, Senior Applications

Consultant at Motorola University, stated that our book "is the best holistic measurement book around." We utilize our World Class measurements to promote continuous improvement at each of our clients. Activity Based Costing is a proven winner. We urge you to initiate a program within your organization today. Today's economic climate has lent a sense of urgency to the elimination of waste and the cost cutting which is occurring everywhere. We must put aside the squabbles and politics so that all of us can work together to change the culture in order to succeed in a World Class environment. Even Vice President Al Gore is attempting to initiate a waste reduction program in the government. Activity Based Costing highlights and identifies the areas of opportunity in government, service and manufacturing. The time to act, the time to begin is the present.

Chapter Ten

IMPLEMENTING ACTIVITY BASED COSTING

An Activity Based Costing (ABC) program does not need to be implemented all at once. In fact, most companies introduce the program into their organizations in a piecemeal fashion. By using this method of implementation, a company can gradually educate its employees about the benefits of the program by showing them actual results. That is why at Pro-Tech, we always recommend that cost management select cost elements which present the largest opportunity for improvement. The next step is to take aggressive steps toward making the costs visible as demonstrated in this book and then taking equally aggressive steps to reduce or eliminate nonvalue-added activities. As the rest of the company becomes excited about the prospect of reducing costs through their own ABC program, cost management can continue to add cost elements until the company is effectively controlling all cost activities within the organization.

Many of Pro-Tech's clients choose to start their ABC programs by selecting Cost of Quality as their pilot project. We agree that this is often a logical place to start with the emphasis throughout the world on quality. However, if other cost factors have a larger effect on organizational results, then start your program in one of these areas. Again, the rule is to select the area with the most potential for improvement.

Selecting an ABC Team
to Develop the Results

For companies who truly want to expand their visibility into cost activities, there is no more effective way than a cross-functional and cross-level team. Figure 10-1 shows one possible combination.

When selecting the six to eight people which will serve on your team, we believe at Pro-Tech that every person who works for your organization is a potential candidate and source of valuable information. We further emphasize time and time again that it is not necessary for all team members to have a financial back-ground. In fact, it is sometimes detrimental for a team to have a traditional cost mindset. As we have demonstrated throughout this book, ABC is anything but traditional. That is why it is often very helpful to have team members who do not have to "unlearn" an old way so that they can grasp the new. Of course, this does not

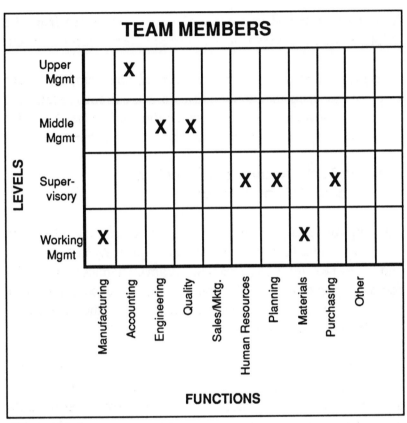

Figure 10-1.

mean that members from accounting or finance should not be on the team. What it does mean is that they should not dominate the team.

Besides being multifunctional, a team should also be multilevel. Every effort should be made to select representatives from each of the organizational levels shown in Figure 10-1. The representative from top management should be a decision maker in the organization. Representatives from working management should come from the employees who work on the floor. Representatives from the middle and supervisory levels should come from the employees who are on the levels between the two just described.

After the team members have been selected, your company's next step is to train them to function as a team. It is extremely important that a team develops a sense of cooperation and ownership early in the process in order to make the most of their time together.

Training Team Members to Function as an Action-Oriented Force

Building ownership in the team's assigned task helps the team focus on results. Team building should begin early in the process and can be best initiated by teaching Team Dynamics to the newly formed group of people. This training should be comprised of a two- to four-hour session. For better absorption of the material, we recommend two two-hour sessions.

One of the first tasks in Team Dynamics is to have the team develop its own mission statement. When a group of people is allowed to undertake this task on its own, we have found that they are far better able to focus their attention on the area which they have been asked to examine. Mission statements should be concise and clear as the example from one of our clients shows:

MISSION STATEMENT

To create an effective cost management tool that determines cost areas of opportunity by exposing all nonvalue-added activities and that provides visibility into the improvement of financial results.

We have also found it helpful to allow teams to create their own names. Again, this will help to develop ownership. Here is a list of some of the more creative names which we have come across at our clients:

TEAM NAMES

- Fasta Pasta
- Cost Cutters
- Profiteers
- Saving Searchers
- Value Adders
- LEAP 2000
- FACE 2000

Lastly, we have also noted in our experience that teams should develop their own set of objectives which they can then use to focus their activities. The only criteria are that the objectives must directly relate to the team's given task. Below is an example of a set of objectives used by a team at one of our clients:

TEAM OBJECTIVES

- TO DEVELOP COST REPORTS THAT REFLECT IMPROVEMENT OPPORTUNITIES THROUGH-OUT THE ORGANIZATION.

- TO INVESTIGATE ALL ACTIVITIES PER-FORMED BY ALL FUNCTIONS AND TO IDEN-TIFY ALL NONVALUE-ADDED TASKS.

- TO DEVELOP A SET OF MEASUREMENTS
  THAT REFLECT CURRENT PERFORMANCE
  AND ESTABLISH GOALS FOR IMPROVEMENT.

- TO RECORD ALL RESULTS FROM PERFOR-
  MANCE IMPROVEMENT ACTIVITIES.

The list of objectives can be as long or short as the team wishes. The emphasis should be put on identifying those activities which are presently causing financial results to be less than the organization would like. Setting objectives, along with developing a mission statement, has been proven effective in providing teams with the foundation for a successful solution to the problems being addressed. The earlier a team is given the chance to start this process of focusing, the more quickly it will produce results.

After training in Team Dynamics, the next training session should be on problem-solving techniques. There are a great many different methods of problem solving being used by companies today, but we have always advocated those methods which truly solve the problem and do not merely attack symptoms. One of the most effective tools to be used by an ABC team is the cause and effect diagram. This tool, often called a fishbone, is primarily used by a team to help them focus. As can be seen in the fishbone completed by one of our clients in Figure 10-2, the diagram allows the team to zero in on a specific function and the activities performed.

The final piece of training for the ABC team should be in activity costing. We always suggest this last segment because most people are only familiar with traditional costing techniques. Team members will need a background in ABC methodology to do their job most effectively.

ACTIVITY FISHBONE

Activities and Opportunities

Administration	Production	Materials
Finance	Labor	Cost of Inventory
Executive	Collection	Handling
	Set-up	Cost of Purchasing
Secretarial	Downtime	Receiving
	Cycle Time	& Shipping
Customer Satisfaction	Cost of Quality	Design Cost
	Delivery	E.C.O.'s
Complaints		Documentation
	Quantity	
Sales	Quality	Engineering

Figure 10-2.

Giving the Team
an Opportunity to be Successful

When the training period is completed, the team now must concentrate on the work at hand and begin to show a return on the investment which the organization has made in their formation. At our clients, we guide the team toward their fishbone and help them select functions and activities within the function to begin their work as a team. Several activities should be selected so as to ensure that as many team members get involved as possible. Teams should be broken down into sub-teams that select an activity to investigate.

These sub-teams should then find the time to visit the assigned function and, with clipboards in hand, trace the activities being performed. We recommend that the team develop a form for use by all members. This form should help members get the information the team needs to capture and costs associated with an activity. Figure 10-3 shows an example of a form developed at one of our clients.

Having captured the needed data, the sub-team can then report back to the ABC team for analysis. The team as a whole will take the data and calculate the cost of people, equipment and idle time. Once this is done, the team can then determine if the activity is value-added or nonvalue-added. In the example shown in Figure 10-3, the cost would be calculated as an average cost for receiving. Then, by determining the number of receipts per day on average, the team can break down the cost into a cost per receipt.

The ABC team is only responsible for highlighting the areas of opportunity and making the organization aware of their analyses. They are also responsible for capturing the savings which occur when improvements are implemented and charting the benefits which will accrue. The implementation of the improvements themselves must be carried out by other teams such as Set-Up Reduction, Supply Management, Quality, Preventive Maintenance, Cycle Time, etc.

Summary

Activity Based Cost reporting should not be a totally new concept to most organizations. Many of our clients tell us that they have been reporting results using ABC methods for a number of years. But, the results they have been reporting have not been used to promote continuous improvement and, for the most part, any such reporting activity has been short lived.

ABC INVESTIGATION FORM

Function: Materials
Activity: Receiving

Task	Time		Equipment	People Involved	Comments
	Work	Sit			
Unload truck	10	10	Forklift	1	Average 50 receipts per day
Get pack slip	5	120	—	1	
Get copy of P.O.	10	30	—	1	Not always available
Verify P.O.	3	240	—	1	
Verify count	15	120	—	1	No counting performed; verify box only
Data entry	10	60	Computer	1	
Move part to inv.	15	240	Forklift	2	Parts sit until Q.C. is finished
File paperwork	5	—	—	1	
TOTAL	73	820			

Figure 10-3.

We strongly believe that ABC should not be just a new way to reflect the results of data gathering. The results should be evaluated by management and used to strategically attack bad performance. In a competitive marketplace, ABC can and will enable management to determine areas of opportunity and to pursue continuous improvement.

By using the techniques presented in this book, your organization will be able to greatly improve its visibility into operational results as reflected by your financial results. ABC has repeatedly shown that it is the ideal method for initiating programs that reduce product cost, increase market share and improve a company's profit picture.

We sincerely hope that you will adopt our techniques and join with our many clients who are enjoying the success brought about by their hard work. **Don't wait to form your ABC team. Just do it!**

Case Study

**MAKE
OR
BUY**

PART 1:

Your company currently manufactures a faceplate used in the assembly of a final product. Using a traditional costing method and the following data, calculate a cost estimate for the faceplate:

MATERIALS

Materials	Cost
Faceplate	$ 4.72
Fan mounting plate	32.41
Front plate	16.54
Back plate	13.85
Standoff	3.46
Heat sink	10.35
Screws (6)	0.17

LABOR

Machine Shop—drilling heat sink (lot of 10 pieces)

Labor rates—Indirect @ $18.14; Direct @ $11.04

Labor Activity	Indirect	Direct
Machine set-up—1.1 hrs		$12.14
Run time—.42 hrs		4.64
Clean and deburr—.25 hrs		2.76
Paperwork—.05 hrs	$0.91	
Inspection—.5 hrs	9.07	
Heat sink anodize to supplier—.5 hrs	9.07	
Heat sink back from supplier—.17 hrs	3.08	

OVERHEAD

Overhead rate = 120% of Total Direct Labor

Now complete the cost estimate.

Material Cost	**$ 81.50**
Labor Cost	**$ 19.54**
Overhead	**$ 23.45**
Total	**$124.49**

PART 2:

Several days after you have completed your traditional cost estimate, you receive a quote that you requested from a supplier who can provide the faceplate for $117.38. At first, it looks like you can save your company $7.11 for each part by buying from the supplier. However, there are additional costs which you will incur if you choose the buy route:

Material	**Cost**
4 bolts for heat sink @ .05 each	$ 0.20

Labor

Bolt heat sink—.95 hrs	10.49
Assembly of faceplate—.06 hrs	.66
Total	$11.35

It actually costs $128.73 to buy the part. With these costs added in, it seems like it costs $4.24 more to buy, rather than make the product. It would seem from this method of estimating that your mind is made up: You should continue making the faceplate. But before you make that decision, you try an Activity Based Cost estimate.

PART 3:

A cost estimate using Activity Based Costing arrives at the following results for overhead, using an indirect labor rate of $18.14. The first set of figures reflects the overhead costs incurred if the company makes the part. The second set of figures reflects the costs incurred if the company buys the faceplate.

Cost of Receiving and Inspection

MAKE FACEPLATE

Parts	Receive Time	Inspect Time
Faceplate	.17 hr	.5 hr
Fan mounting plate	.17 hr	.5 hr
Front plate	.17 hr	.5 hr
Back plate	.17 hr	.5 hr
Standoff	.17 hr	.25 hr
Heat sink ext.	.17 hr	.5 hr
Totals	**1.02 hrs**	**2.75 hrs**
Cost	**3.77 hrs X $18.14 =**	**$68.39**

BUY FACEPLATE

Parts	Receive Time	Inspect Time
Faceplate	.17 hr	.5 hr
Heat sink	.17 hr	.5 hr
Totals	**.34 hrs**	**1.0 hrs**
Cost	**1.34 hrs X $18.14 =**	**$24.30**

The company would spend **$44.09** less on Receiving and Inspection if it bought the faceplate instead of making it.

Cost of Stocking

MAKE FACEPLATE

Parts	Time to Stock	Time to Pull Job
Faceplate	.03 hr	.03 hr
Fan mounting plate	.03 hr	.03 hr
Front plate	.03 hr	.03 hr
Back plate	.03 hr	.03 hr
Heat sink drilling	.03 hr	.03 hr
W/O for EPR	.17 hr	.03 hr
Totals	**.32 hrs**	**.18 hrs**
Cost	**.50 hrs X $18.14 =**	**$9.07**

BUY FACEPLATE

Parts	Time to Stock	Time to Pull Job
Faceplate	.03 hr	.03 hr
Heat sink	.03 hr	.03 hr
Totals	**.06 hrs**	**.06 hrs**
Cost	**.12 hrs X $18.14 =**	**$2.18**

The company would spend **$6.89** less on Stocking if it bought the faceplate instead of making it.

Cost of Buying

MAKE FACEPLATE

Parts	Parts Order Time
Faceplate	.1 hr
Fan mounting plate	.1 hr
Front plate	.1 hr
Back plate	.1 hr
Standoff	.1 hr
Heat sink ext.	.1 hr
Totals	**.6 hrs**
Cost	**.6 hrs X $18.14 = $10.88**

BUY FACEPLATE

Parts	Parts Order Time	
Faceplate	.1 hr	
Bolt on sink	.1 hr	
Totals	**.2 hrs**	
Cost	**.2 hrs X $18.14 =**	**$3.63**

The company would spend **$7.25** less on Buying if it bought the faceplate instead of making it.

Cost of Planning

MAKE FACEPLATE

Parts	W/O Time
A—W/O cut and drill heat sink	.03 hr
B—WIP receipt Machine shop	.03 hr
C—W/O send to EPR	.17 hr
D—Heat sink to EPR	.5 hr
E—Pick-up heat sink from EPR; Cut receiver	.17 hr
Totals	**.90 hrs**
Cost	**.90 hrs X $18.14 =** **$16.33**

BUY FACEPLATE

Parts	W/O Time
W/O assemble faceplate	.03 hr
Totals	**.03 hrs**

Cost   .03 hrs X $18.14 =   **$0.54**

The company would spend **$15.79** less on Planning if it bought the faceplate instead of making it.

Now, let's summarize what we have calculated and determine the Total Cost for the faceplate if we buy it and if we make it:

	Make	**Buy**	**Difference**
Material Cost	$ 81.50	$117.38	$(35.88)
Labor Cost	19.54	11.15	8.39
Overhead Cost			
Receiving	68.39	24.30	44.09
Stocking	9.07	2.18	6.89
Buying	10.88	3.63	7.25
Planning	16.33	0.54	15.79
Totals	$104.67	$ 30.65	$ 74.02
Total Cost	$205.71	$159.18	$ 46.53

Now, as you can see, it actually costs **$46.53** less to buy the faceplate than to make it, even though a traditional costing method indicated the opposite. It pays to use Activity Based Costing because it reflects the true total cost of a part. How many other make/buy decisions do you make in your company based on traditional methods?

Bibliography

Activity Based Costing, course workbook, PT Publications, Palm Beach Gardens, FL, 1992.

Advanced Level Manufacturing Cost Estimating, course workbook, Society of Manufacturing Engineers, Dearborn, MI, and PT Publications, Palm Beach Gardens, FL, 1992.

Accounting Principles, Roger H. Hermanson, James Don Edwards, and R.F. Salmonson; Business Publications, Inc., Plano, TX, 1987.

Megatrends 2000, John Naisbitt and Patricia Aburdene; Avon Books, New York, NY, 1990.

Accounting and Management: *Field Study Perspectives*, William J. Bruns, Jr. and Robert S. Kaplan, Editors; Harvard Business School Press, Boston, MA, 1987.

Cost Management for Today's Advanced Manufacturing, Callie Berliner and James A. Brimson, Editors; Harvard Business School Press, Boston, MA, 1988.

How to Implement Kanban for American Industry, Raymond S. Louis; Productivity Press, Cambridge, MA and Norwalk, CT, 1992.

Beyond the Bottom Line, Carol J. McNair, William Mosconi and Thomas Norris; Dow Jones-Irwin, Homewood, IL, 1989.

Cost Management in the New Manufacturing Age, Yasuhiro Monden; Productivity Press, Cambridge, MA and Norwalk, CT, 1992.

TQC for Accounting, Takashi Kanatsu; Productivity Press, Cambridge, MA and Norwalk, CT, 1990.

Fortune, Marshall Loeb, Editor; James B. Hayes, Publisher; New York, NY.

Set-Up Reduction: *Saving Dollars with Common Sense*, Jerry W. Claunch and Philip D. Stang; PT Publications, Palm Beach Gardens, FL, 1989.

Wall Street Journal, Robert Bartley, Editor; Peter R. Kann, Publisher, New York, NY.

Harvard Business Review, Theodore Levitt, Editor; James A. McGowan, Publisher, Boston, MA.

Industry Week, Charles R. Day, Jr., Editor; Vincent A. Castell, Publisher, Cleveland, OH.

Managing Automation, Robert Malone, Editor; Ralph E. Richardson, Publisher, New York, NY.

USA Today, John C. Quinn, Editor; Cathleen Black, Publisher, Washington, DC.

Made In America: *The Total Business Concept*, Peter L. Grieco, Jr. and Michael W. Gozzo; PT Publications, Palm Beach Gardens, FL, 1987.

World Class: *Measuring Its Achievement*, Peter L. Grieco, Jr.; PT Publications, Palm Beach Gardens, FL, 1990.

Behind Bars: *Bar Coding Principles and Applications*, Peter L. Grieco, Jr., Michael W. Gozzo, C.J. (Chip) Long; PT Publications, Palm Beach Gardens, FL, 1989.

Just-In-Time Purchasing: *In Pursuit of Excellence*, Peter L. Grieco, Jr., Michael W. Gozzo, Jerry W. Claunch; PT Publications, Palm Beach Gardens, FL, 1988.

The World of Negotiations: *Never Being a Loser*, Peter L. Grieco, Jr. and Paul G. Hine; PT Publications, Palm Beach Gardens, FL, 1991.

Supplier Certification II: *A Handbook for Achieving Excellence through Continuous Improvement*, Peter L. Grieco, Jr.; PT Publications, Palm Beach Gardens, FL, 1988.

Index